Sosthène Trésor N'NANG EBANE

AN OBSERVER'S JOURNEY INTO THE HEART OF UNIVERSES

Sosthène Trésor N'NANG EBANE

AN OBSERVER'S JOURNEY INTO THE HEART OF UNIVERSES

Mathematics, Spiritual Arts, Jews and Calendar

ScienciaScripts

Imprint

Any brand names and product names mentioned in this book are subject to trademark, brand or patent protection and are trademarks or registered trademarks of their respective holders. The use of brand names, product names, common names, trade names, product descriptions etc. even without a particular marking in this work is in no way to be construed to mean that such names may be regarded as unrestricted in respect of trademark and brand protection legislation and could thus be used by anyone.

Cover image: www.ingimage.com

This book is a translation from the original published under ISBN 978-3-8416-3595-2.

Publisher:
Sciencia Scripts
is a trademark of
Dodo Books Indian Ocean Ltd. and OmniScriptum S.R.L publishing group

120 High Road, East Finchley, London, N2 9ED, United Kingdom
Str. Armeneasca 28/1, office 1, Chisinau MD-2012, Republic of Moldova, Europe
Printed at: see last page
ISBN: 978-620-7-70542-9

Contents

Fig: Dz^ Ayem
~<(o)>~
(Unitary, binary and trinitary symbols<183 +1>)
GENESIS AND EBANETH DESIGN (LCS4)
nnangebanes@gmail.com
N'NANG EBANE Sosthene Tresor

*
**This book
is dedicated to the memory of
Mr EBANE MINTSA Jean Paul
(1954-2001)**
*** * ***

(National Gendarmerie)

INTRODUCTION

*"I like to think that many young people from your ðënëzauon will have the vocation to seek out at the end of their university ëtudes to uncover the Ir'sors of our culture. They are still buried in the humus of our villages "*It is generally said of the African continent to be built around a close relationship between the old and new generations, based mainly on the sharing of knowledge, in order to take in turn the queens of power. This heritage of knowledge is most often passed on through initiation rites, without which it would be impossible to be accepted within the group as an authority figure. The invisible world is thus granted a certain power, without which it would be impossible to govern carnal beings. During these initiatory spiritual celebrations, k

traditional instruments such as the Sacred Harp, the Mvet Ekang, and more, to establish the spirit of justice in human nature in order to silence its yearning for self-satisfaction at the expense of the group called upon to remain under its care. Religious tools emit melodious sounds with the aim of connecting the human spirit to its universe of essence, while incorporating within them geometric forms sometimes relating to the organisation of the universe, putting vibration first and foremost as the very expression of life. Geometrical and spirituality combine in the integration of the human soul into its universe of essence, giving rise to an entirely different appreciation to this new vision: *'The term gëomëtrie sacrie is ëgalement nИзë to indicate the application of gëomëtrie to a religion and to all exotërism as a direct consëquence of the conception dëcrite below of the cosmos. Gëomëtric forms are used in all cultures, in the construction and structuring of sacred buildings such as temples, mosques, mëgaliths, churches and sacred places such as altars and tabernacles, as well as in the criation d'art sacri "*One of the particularities observable on these spiritual instruments is the incorporation of a so-called flexible part within another called rigid, that is to say practical works of two contradictory realities. This contradictory duality can also be seen in the dry and rainy seasons. The first refers to a static action, while the other defines an action in progress and sound. The periods follow one another in succession, the whole referring to a cycle that renews itself continuously, without the intervention of mankind, undoubtedly giving a whole new particular flavour to the expression of beauty: *"Art in all its forms is a means of expressing civilisations. It is the mark of the culture, social, religious, political and economic organisation of a people at a particular moment in its history. In a context of ritual art, it enables people to act and live in harmony with their environment. In this sense, masks reflect the whole of the civilisations in which they are produced, understood and appreciated.* The use of art as a means of expression is most often subject to a palpable and pocessive reduction of the organisation of universal realities to a representation that can be manipulated by mankind. It is the process of human appropriation of the planet in its true dimension, as a being receptive to the work of the creator. Art is a practical rendering of the unfolding of a history experienced through the eye of the observer, realised on a palpable and pocessive medium, in order to pass it on to future generations. It is part of the unconditional preservation of the history of a given civilisation. Although ritual art relates to a distant common past, it is fiercely exclusive towards certain members of the same ethnic community, on the grounds of strict respect for ancestral rules. Observation of their different skeletons seems to plunge these ritual arts into a universe with strong mathematical connotations, tending to remove them from their initial comfort zones. That guided by a curiosity about the quest for knowledge of civilisational memories, which can contribute to a lesser extent to a more complete understanding of them. The one built around the hypnotising and intoxicating mathematical organisations of their multiple skeletons: *"lallK'malics: science that ëtudies, by means of dëductive reasoning, the propriëtës of abstract beings (numbers, gëomëtric figures, functions, spaces, etc.) as well as the relationships between them.*Colonisation is a period of history that most often tends to recognise Africans as an integral part of the human revolution along Western lines, even though it was more remote at the time. The history of Pharaonic Egypt, which is still the subject of debate today, concerning the exact recognition of the peoples who lived there during this glorious period, unfortunately cannot be of much help in understanding these ritual arts. Generally speaking, the ritual arts we have identified feature the pyramid in one way or another as their main framework. The particular configurations of this geometric figure, a key element in the expression of spiritual greatness within each object of study, tend to define their holders as possessors of unique knowledge. In the distant past, the geometric symbol seems to have been used unanimously by all these different peoples. It is only natural that each civilisational memory should be discovered in turn, with its own particularity relating to the Egyptian monument. The *Ngombi is a plucked string musical instrument from Central Africa (Central African Rëpublique, Rëpublique du Congo and Gabon). It is the arqiK'e cirqiK'e harp used by the KwAlA (or Kele), the Fang and the Mbochi.* "The Ngomba comprises a diagonal triangle of 8 cordophones in a parallel and fractal posture, extracted from a square measuring 9 cm in length. It is constructed so that the length is one unit greater than its main diagonal, which is symbolised by the number 8, which in turn knows the number 7 as an internal numerical value. The triplet 987 implies that the largest number is incompatible with the smallest value, and so on. Next: *"The Mvett, sometimes Acrit Mvet, or Mver, which refers to a stringed musical instrument spelt with a capital letter. The term dëfinit un ensemble de lAcits guerriers qui se joue accompaònë du dit instrument. Le Mvet designe a la fois la harpe- cithare ainsi que les recits heroiques que declame un barde appellee mbom-mvett (joueur/ conteur de Mvett)* " In contrast to the above, the instrument consists of a fractal diagonal triangle with a bridge forming a right angle with its base. It presents a manifest arithmetical triplet according to its construction, offering both the possibility of ascent and regression. Note: T(n)=n+1+n equals T (3)=3+1+3=7 and T(n)=n+(n+1)+(n), equals T (3)=3+(3+1)+3=3+4+3=10, the whole revealing the image of an isosceles triangle. The two ritual objects are held in part by the following ethnic group, the majority of whom live in the heart of the equatorial forest: *"The Fang, whose real ethnic name is thought by some to be M'fang, are a Bantu ethnic group found today in Central Africa, mainly in southern Cameroon, Equatorial Guinea and Gabon, but also in the Republic of Congo, the Central African Republic and Sao Tome and Principe. The enigmatic face of the mask is evenly triangular. Under the closed, almond-shaped eyes, as if swollen by sleep, the high cheekbones become rounded (....). The most common pattern, in the shape of scales, comprises nine lozenges"*. The mask patterns correspond to the transformation of the square into a rectangle, in order to emphasise its practical unity with the trilateral figures. The most common pattern is that of a square joined to four triangles. The triplet formulas listed above are used to achieve this transformation. The ritual objects in question are specific to a particular ethnic group: *"The Punu (also known as Bapunu, plural of Mpunu) are an ethnic group mainly found in southern Gabon. The Punu migrated to southern Gabon (in the Ngounie basin) in the 18th century"*. *The Fang and the Punu are two peoples living in Gabon who share no affinities according to their respective migratory histories, yet the ritual arts they possess surprisingly share some very striking common characteristics linked to the science of matl'K'matiques, of which the pyramid is a symbol.* By extension: *"Songo is an African strategy game from the sowing game family. There are several variants, depending on the geographical region, which take on other names. The name Songo is typical of the inhabitants of Central Africa (Cameroon, Gabon, Equatorial Guinea). In West Africa, more specifically in the Ivory Coast and Niger, it is called Awate or Aweri, while in Benin it is called Adjito and jeu de six in Togo.* As for the game of sowing, it is drawn from the diagonals of a pyramid. It is organised around the number 7, which this time is associated with the number 9 because it is part of the numeric triplet family. The layout of the deck between the players facing each other is a practical perception of the axes of symmetry. The triangular and diagonal stagnation of the 9 cm long square is 28, which is double the number 14 that makes up the number of squares. The addition by singular numerical decomposition of the number is equal to this same number. Count   1,2,3,4,5,6,7   and   add   them   together $(1+2)+3+(4+5)+(6+7)=(3+3)+(9+13)=6+22=28$. In the same vein, the Jewish candlestick also has 7 branches: *"The Menorah is made up of the prifix "më" indicating the origin of an object associated with the hëbraic root "norah", "nourah", from "nour", nor "flamme" to the fëminin. Menorah therefore means from "the flame", "which comes from the flame". According to the Kabbalah, this flame is none other than the shëkina or "prisence of God".* There are two Hebraic candlesticks, one with 7 branches and the other with 9 branches called the Hanukkiah. Both are a perception of a diagonal and fractal triangle. They constitute the 3rd and 4th arithmetical triplets in succession: $3+1+3=7$ and $3+4+3=10$ (7) , $4+1+4=9$ and $4+5+4=13$. According to Maimonides, the branches of the Menorah are similar to the letter Y, a typical variant of the branches of the Upper Nile. The Egyptian context always comes into play in one way or another when referring to the skeletons of the sacred arts. In a typically Hebraic context, the luminous array of the Menorah reveals the number 27 as the result of its numerical condensation, which is exactly the same number of letters that make up the Hebraic alphabet. In the same perspective, the number 9, by adding the results of its arithmetical triplets, relates to the number 22=9+13, relating to the same alphabet. They sometimes appear with all the branches of the same size, and sometimes with the main branches higher than the others, corresponding to the possibility of ascent and regression expressed by the triangular and fractal diagonometric cordophones of plucked string instruments. The growth and decay of the branches can refer to both the rainy and dry seasons that can be identified within the material reading of time. Finally, *"The term calendar, which comes from the Latin word calendrium and means 'account book' dësignifies a document containing a long list of information: days, weeks, months of a given year. Definition: this is a system for dividing time to make it easier to identify dates. "*The number 7, through the 31-day months, will form a temporal axis with superimposed and inferior undulatory variations, so that we can see that the calendar is a summation of the 13-branch candlestick and the 9-branch candlestick. As a bonus, the Menorah will be discovered as an integral part of the 13-branch candlestick. In the end, the calendar will emerge as the nerve centre of all the ritual objects listed. By looking at all these spiritual arts in the light of the breve apergu described above, it seems possible to provide additional information about how they are understood by the community. It would not be wrong to think that multiplicity is an exhaustive view of unity, which has vanished as a result of the fierce quest for power, based on the shameful feeling of racial segregation.

I-) THE NGOMBI

https://www.bantoozone.org>ngoma-harpe-fang-music-gabon

*

A souvenir
elogious about
Mr MINTSA MI EBANE Thomas Leonce
((01/01/1988)- (14/03/2019)

*** * ***

Master II, History and Archaeology

1-) The nature of the triangle

An uninitiated observer of traditional religious instruments embarks on a race to find mathematical codes, initially within the Sacred Harp. *"The Ngombi is a plucked string musical instrument from Central Africa (Central African Rëpublique, Rëpublique du Congo and Gabon). This is the Harpe arqiK'e usedëe by the K\\'ëlë, (or Kele), the Fang as well as the Mbochi"* Looking at this paragraph describing the Harpe arquee, we realise that it is held by several African peoples, who unfortunately no longer share a common homeland, but possess a single history thanks to this civilisational memory. This is how art comes to be seen as a piece of identity shared by its different peoples. The unconditional preservation of the history of peoples seems to have been a matter of great importance since time immemorial. The idea of safeguarding implies another, relating to the recipient, who in turn has a duty to perpetuate it. However, can this continuity always be pursued under the watchful eye of the initiatory order in the strict sense of the term?

The study will focus essentially on the link between the neck, the cordophones, and the trunk, the civilisational memory. The Sacred Harp has 8 cordophones. However, they are arranged from top to bottom in ascending order, while at the same time the descending order is revealed by starting at the lowest level, with the longest stringophone moving towards the shortest at the top. The cordophones therefore design a unique element that has both the possibility of reaching for the heavens and the possibility of reaching downwards, two opposite sides of the same coin.

a-) the isocele triangle

(Unconditional reminder of the petrol order)

The sculpture has exactly 8 perforations on its neck and 8 perforations on its trunk, which are joined by 8 cordophones in the middle. In view of the above, since the number of upper perforations is exactly the same on the lower level, we can deduce that the triangle revealed there is isolated:

"Among the trilateralëres figures, the isocele triangle is the one that has only two ëgal sides".

In this configuration, the memche is taken to symbolise a side measuring 8 cm, as is the trunk, from which the longest of the cordophones indicates its base, which of course has a particular length. The number of cordophones is not taken into account in this case, giving rise to a first hypothesis.

The isosceles triangle itself evokes the harmony of opposites, through its two sides of equal size with the neutral triangle. Neutrality implies the presence of an additional, opposite element within the identical linear pair. In a certain dimension, it can describe the threefold view of human life: mind-matter-spirit. Indeed, it is said that the human being is in essence spirit, so his body is merely an expression conjugated with it. However, death, which separates the spirit from the body, returns the spirit to its essence. In the light of the above, the isocele triangle evokes

9

not only the spirit's vain passage through matter, but also its return to its state of nature. In simple terms, it momentarily binds matter to the spirit, only to return it later - a temporary rest. It is humanity's unconditional return to its essential being.

b-) the equilateral triangle

(State of totalitarian dependence)

As a second hypothesis, it is also possible to consider the triplet perception of the number 8 within the trilateral figure found in ritual art. It has 8 upper perforations, 8 cordophones in the centre, and 8 lower perforations. From this point of view, the number 8 will therefore be considered as the measure of each of the lengths of the three (3) sides of the triangle, from which we obtain an equilateral triangle.

"Among the trilatëres figures, the ëquilatëral triangle is the one that has its three côtës ëðaux".

The triangle, with its sides of the same length, can be related to the idea of perfect balance, or even better, to the harmonisation of similar values that can be seen in different places. Indeed, although each side shares a common length, their postures are quite different. The base and the two sides at the top seem to define a permanent cycle, which is why the equilateral triangle can sometimes be mistaken for a circle. It is an energy that constantly renews itself, and continues to do so until infinity. It is no more and no less than the expression of a communal reunion of elements of the same nature. In a more extreme dimension, the equilateral triangle evokes the annihilation of the hold of matter over the mind. The presence of a certain spiritual entity leads the human soul to seek refuge in other heavens, against its will. The trilateral figure describes the hypnotising state of the spiritual work on the human dimension. It is the state of totalitarian dependence of human nature.

c-) isosceles triangle and equilateral triangle

(Degrowth and growth)

The hypotheses put forward above concerning the precise nature of the triangle found in Ngoma suggest that the isosceles triangle is different from the equilateral triangle, since it incorporates a side (1) of different size from the other two (2) with which it is associated, which on the contrary are of the same length. It is important to note, however, that the transition from the isosceles triangle to the equilateral triangle is possible by enlarging the neutral side, so that it can merge with the other two (similar sides). It is important to understand that size and smallness are generally two adjacent limits linked to the evolution of a given element. The waters of a river know how to rise and fall, to justify the manifest view of geometric shapes on nature. Geometric figures offer the possibility of ascent and regression, like the cordophones of sacred art, a single element indicating both height and base. The mountain is linked to both earth and sky, as a physical element of nature.

This observation leads us to define two mathematical formulae for the different trilateral figures. $T(n)=n+1+n=(2xn)+1$ equals $T(3)=3+1+3=(2x3)+1=1+6=7$ and $T(n)=n+n+n=3xn$ equals $T(3)=3+3+3=3x3=9$.

2-) The nature of quadrilaterals

10

There is no shadow of a doubt that the trilateral figure is present in the framework of the Sacred Harp in two ways. The first is built around the triangular nature of its morphological organisation, while the second relates to the number 8 appearing 3 times, i.e. 8 upper perforations, 8 cordophones, and 8 lower perforations. However, we are only going to focus on the numerical aspect as far as the quadrilaterals are concerned, i.e. the square on the one hand and the rectangle on the other.

a-) the rectangle

(transfer of celestial and terrestrial forces)

It should be pointed out in passing that the Ngombi also includes 8 vices, used to adjust the unique sound of the cordophones. So now we have 8 valves, 8 upper perforations, 8 cordophones in the centre, and 8 lower perforations, going from 888 to 8888. In view of the above, there is just one number 8 added to the triplet 888, just as a side of the same length is added to the two others already of the same length within an isosceles triangle to give rise to an equilateral triangle. To bring this out, it's a good idea to place 8 points on the horizontal plane, respecting the layout of the elements listed above on ritual art, which gives us the following sketch:

Fig: 1'equisse rectangulaire

Δ Δ Δ Δ Δ Δ Δ Δ

Δ Δ Δ Δ Δ Δ Δ Δ

Δ Δ Δ Δ Δ Δ Δ Δ

Δ Δ Δ Δ Δ Δ Δ Δ

Source: N'NANG EBANE Sosthene Tresor

From the above, we can easily determine the length of the rectangular sketch, which is 8 cm, while the width is 4 cm. The quadrilateral thus offers a numerical view of the length and width, i.e. 8 by 8 and 4 by 4. The pair (8;8) may designate the first axis of symmetry, while the second pair (4;4) refers to the second axis of symmetry:

"A rectangle is a quadrilateral with four right angles.
- *A rectangle is a paralldlogram*
- *The rectangle has parallel opposite sides of equal length*
- *The rectangle has two axes of symmetry: the mëdians of its sides*
- *In a rectangle, the diagonals have the same midpoint and the same length.*
- *The midpoint of the diagonals is the centre of symmetry of the rectangle".*

The properties listed above fit perfectly with the rectangular sketch obtained through the horizontal alignment of the number 8, which is repeated in the Sacred Harp. The quadrilateral reconciles parallelism and perpendicularity, expressing the harmony of opposites. Above all, it evokes the unity of celestial and terrestrial forces. The vertical sides are shorter than the horizontal ones, to signify two practical facts. The celestial forces descend to earth in a material form, just as the

11

terrestrial forces rise towards the heavens according to the work of the spirit. The widths thus symbolise the zones of metamorphosis, of each opposing energy. Rain falls from the heavens to the delight of earthly beings, who in turn repay it by rising towards the heavens. The rectangle expresses the idea that every manifestation always has consequences. It's important to understand that my human life takes place within a specific framework, but outside it there lives a vibratory energy that allows the internal elements to move. The idea that emerges here is that God cannot live on earth with humanity, yet humanity evolves under his protective gaze.

b-) the square

Both the square and the rectangle are built around the number 8888, which refers to the number of sides they have. To discover this, we naturally need to reconsider our rectangular sketch above.

Fig: l'equisse rectangulaire

Δ   Δ   Δ   Δ   Δ   Δ   Δ   Δ

Δ   Δ   Δ   Δ   Δ   Δ   Δ   Δ

Δ   Δ   Δ   Δ   Δ   Δ   Δ   Δ

Δ   Δ   Δ   Δ   Δ   Δ   Δ   Δ

Source: N'NANG EBANE Sosthene Tresor

A close look at this sketch reveals that it is actually made up of two (2) blocks of 16 triangles each, with two (2) squares revealed within the rectangle. It is right to separate them as shown in the sketch below:

Fig: 1'equisse rectangulaire

In the light of the foregoing, it is clear that the rectangle is indeed full of two squares. The square is thus seen as a component of the rectangle, in a duplicated perception. Duality is thus revealed as a multiple perception of unity. It is very important to understand up to this point in the analysis that the number 3 is obviously integrated with the number 4, symbolising the triangle on the one hand and the square on the other.

Both the square and the equilateral triangle relate to the idea of perfect unity, to the cyclical view of an energy that is constantly renewing itself. In a horizontal posture, it evokes typically earthly facts, while in a vertical posture, it gives an account of celestial realities. In the latter posture, only one right angle touches the ground, while the other three seem to reach for the heavens. The significant and true celestial value is naturally built around the triangle, which is part of its internal component. The trilateral figure is present in both the rectangle and the square. It is also their common denominator, like the 4 angles and 4 sides they have. This is why the number 8 is a numerical symbol for quadrilaterals. But the triangle gives a celestial character to the quadrilateral.

c-) the rectangle and the square
(condensed and exploded forms)

It is clear to see that the rectangle is larger than the square, which offers a more restricted view. So, moving from the rectangle to the square relates to the idea of decline, whereas moving from the square to the rectangle expresses growth. The fact that the square is inscribed within the rectangle implies that growth and decay are two evolutions of a single object. This vision of growth and decay at the same time reveals the image of the triangle as a figure that is mostly very relevant in expressing size and scale, which corresponds very exactly to the organisation of the cordophones of the Sacred Harp. This is no surprise, given that the trilateral figure is a component of the quadrilaterals. So we have a vision of the forms of the rectangle that is both condensed and fragmented. A single element that can both unite and disunite. The sun alone has three very distinct phases: sunrise, zenith and sunset, even though it is a unique element in nature. What can we say about the moon and its different phases? The idea is that natural elements can sometimes go haywire, as in the case of the solar eclipse, when the moon seems to swallow the sun, only to spit it back out later. And yet, there are two specific periods during which each element should really have its moment of glory, but the fact remains that the moon can erupt during the day. The square and the rectangle are two distinct quadrilaterals, but they are capable of uniting under a single form that is the first. One unconditionally implies the presence of the other. Just as the equilateral triangle carries within it the isosceles triangle. In the light of the above, observation tends to reveal a kind of kinship between the geometric figures, accompanied by both ascent and regression, justifying the triangle as a highly characteristic example of this contrasting evolution.

3-) the fractal configuration of the triangle in numbers
(gradual diagometric evolution)
a-) the configuration of the number 8

It would be incomprehensible for some to understand how the number 7 is expressed within the triangle of the Sacred Harp instrument, even though it is made up of 8 cordophones. It should be remembered that we are merely allowing ourselves to be carried away by the fantastic atmosphere offered by all these religious arts, and by the Ngombi in particular. The numerical reading in relation to the triangle and the rectangle on the object of study has effectively enabled a passage from the number 888 to the number 8888. In the light of the above, the triangle offers a threefold view of the number 8, while the square offers a fourfold perception of the number 8, from which the number 7 is obviously obtained by simple addition. There are in fact two orders of evolution in relation to what has just been said: the number 7, which is one unit smaller than the number 8, is perceived via the triplet and quadruple addition of the second. Also, the passage from the number 24=3x8 to the number 32=4x8 shows that the number 8 has been added. In other words, the simple number 8 evolves gradually within the ritual art, starting from itself: we have: 8+(8+8)+(8+8+8)+(8+8+8+8)=8+16+24+32. If we fill in each number 8 with its numerical frequency of appearance, we get the following numbers: 1+2+3+4=(1+2)+(3+4)=3+7=10. We can see that the number 8 is multiplied by a different number in an order of growth known as a fractal. The fractal configuration: *"An object whose structure is ƶёρёκ' at all scales of observation"* of the trilateral figure leads us to take the above as an illustration:

Fig 3: Mummies and pyramid

Sources: https//:www.mummies2pyramid.com

The present triangle offers exactly a gradual evolution from the inside to the outside, just as from another point of view, we can obviously observe a decay from the outside to the inside, exactly like the cordophones of the traditional Sacred Harp instrument. The triplet 123 is half of the triplet 345, because if we assume:

1+2+3=6 and 3+4+5=12, then the external value is 6 times greater than the internal one. In this case, we are talking about the triplet perception of the number 8, which is double the number 12. We get: 6+(6+6)+(6+6+6)+(6+6+6+6)=6+12+18+24.

b-) Diagonals and parallelism

It is now appropriate to start with the triplet 123 and work our way up to the numbers 888 and 8888, which appear in the sacred art plan. It is important to understand, however, that each number in the triplet must occupy an avant-garde posture, so that they can be assimilated to an identical number, seen in triplet form.

We pose:(x

123+132= 255

213+231=444

312+321= 366

We can see that the numbers 255 and 366 are different from the number 444, which has three identical digits. So, to make these numbers uniform, we need to invert the data from the diagonometric form (x) to the parallel form (=).

Let: (=)

123+321=444

213+231=444

312+132=444

Considering first the number 888,

The equation is: 444+444=888

The number 888, symbolising the triangle through the algebraic reading of the Ngoma, is exactly twice the number 12, which can be seen in fractal form on the Egyptian triangle illustrated above. We can go further, using the number 8888 as a starting point.

Finally, let's consider the number 8888

Let: (x)

4123+4132=8255

4213+4231=8444

4312+4321=8633

Once again, we need to invert the X data as in the previous case:

Let: (=)

4123+4321=8444

4213+4231=8444

4312+4132=8444

Finally, the previous data must be added to the current data to obtain the instrument numbers.

Let: (=)

8444+444=8888

8444+444=8888

8444+444=8888

The sacred art consists firstly of 8 upper perforations, 8 plucked strings in the centre, and 8 lower perforations, symbolically revealing the number 3. Secondly, by adding the 8 upper vices used to make each cordophone more efficient, we obtain the number 8888, revealing the number 4. In view of the above, the number 7 is cleverly hidden within the Ngombi with 8 cordophones.

Let's consider once again the following addition: 8444+444, by counting each digit singly or filling them with the digit 1, we obtain the digit 7 by addition, i.e. 1111+111=7. If we also consider the number 8888, we can first have the number 888 (perforations and cordophones), then the number 8888 (vices, perforations and cordophones), i.e. 3+4=7.

It is possible to go even further by involving the notion of numerical or arithmetical stagnation through 8 by 4. In other words, the number 4 serves as a vector for arithmetical assimilation within the number 8, through its increasing fragmented form.

We pose:

8stg(4)=12345678

=2+(1+3)+4+(5-1)+(6-2)+(7-3)+(8-4)

=2+4+4+4+4+4+4

=2+(4x6)

=2+24

8stg(4)=26

If we consider the numeric condense **2+4+4+4+4+4+4**, filling each number with 1, we obtain the number 7 once again: **1+1+1+1+1+1+1=7.**

In an equilateral triangle, lines can be drawn parallel to its base and height in a fragmented order, the number of which is a small unit less than the length of its sides. The side of the Ngoma triangle measures 8 perforations/cm, so 7 lines can be drawn parallel to its base and height.

Cordophones always reflect the same implacable reality according to which everything necessarily rests on a given foundation, where each element of the universe has a certain origin. The relationship between the number 8, which refers to the value outside the triangle, and the number 7, which designates the value inside, shows beyond any doubt that it is absolutely necessary to reach 7 before 8. So, from this perspective, the fact that new generations learn from the old is a mathematical logic revealed by the cordophone triangle of sacred art.

From this perspective, the triangle itself evokes both ascent and regression, the ability to return to an initial stage, from which the end of the story can be summed up in the idea of revolution: a return to square one.

The identical tripet 444 in fact symbolises the original energy to which the other energies must be assimilated, which is why the numbers 255 and 366, by inverting their X numbers, end up with the same result. So, the departure of the numbers from 255, 444, 366 to 444, 444, 444, shows of course that the cycle of a given element ends up with it recovering its initial form.

Man is a spirit, living in a body for a specific period of time before becoming a spirit again. The two (2) universes need to be filled in here. The triplet shows that there is a neutral energy that unifies the two opposing realities, serving as a passage between them.

The number 12 refers to the four triangles associated with their square base, while the number 7 simply refers to the unity of the square with the triangle. In this way, the pyramid is presented in its complete form on the one hand, and in half on the other. This is how we speak of a single element with countless facets.

The arithmetical triplets offer a very particular perception of the triangle contained within African sacred art. In effect, they imply the perfect harmonisation of numerical values, defining the trilateral figure as a symbol of purity. The numbers are assimilated to refer to a single number with triplet components of the same value, albeit of different basic values. The observation made about numbers relates to a major idea: multiplicity is combined within a completely homogeneous unit. Humanity is defined not only by its morphological and linguistic characteristics and its geographical names, but also by its ability to solicit the contribution of its fellow human beings in order to achieve a given project. The project in question will be a unique work of art with two non-perceptible contributions in terms of its functioning and its usefulness.

Another very important fact is the mobility of arithmetic values in order to obtain an identical and generalised number. In fact, the numbers move from a diagonometric position (x) to a so-called parallel position (=), in order to express tripartite or quadruple homogeneity. If the triplets refer to the triangle, and the numbers refer not only to the diagonals, but also to the parallel lines, then the trilateral figure is associated with a quadrilateral. A purely algebraic analysis reveals the unity of the triangle with the quadrilateral, whose numerical symbol is the number 7.

Let's go back to the image used as an illustration. The idea of the Ngombi civilisational memory as originating in Pharaonic Egypt may not be very convincing. However, we can admit that the fractal perception of the trilateral figure fits perfectly with the image proposed above. The numbers evolve according to the gradual order of a figure taken as a construction vector until its completion. What's more, the diagonal and parallel view of numbers by addition relates to a quadrilateral. Since it is a single result which is generalised into both three and four possibilities, then it can only be neither more nor less than the unity of the triangle with the square, hence the pyramid. Note that the mummy's arms are crossed over his chest in imitation of the diagonals, which are made practical through the coloured numbers of our algebraic analysis. What about the triplet 888 referring to the number 24, half of which is shown in the image of the mummy?

c-) the natural order of the numerical triplet (5)

Let's look at the order of growth of the digits and numbers from 1 to 10, in three distinct categories:

We have 12345678910

18

Let's identify the even numbers: 246810 (A)

Let's remember the other natural one: 12345678910 (B)

Let's draw out the odd numbers: 13579 (C)

Let's determine the growth and regression orders:

(A): -2 and +2

(B): -1 and +1

(C): -2 and +2

Considering the three sets (A), (B), and (C), we have : (A)=(C) and (C)=(A)

(A)#(B) and (C)#(B)

In view of the above, the standard triplet for the revolution of numbers is: -2-1-2=-5 or 2+1+2=5. It is essential to recognise that the arithmetic triplet is mostly odd.

4-) the diagonometric nature of the triangle

a-) diagonals

Throughout the previous section it was shown that the arithmetical triplets relate to the triangle as well as the quadrilateral. However, this analysis was based on simple numerical observation, and did not make it possible to determine whether the square or the rectangle was really associated with the tripartite figure. To discover the true identity of the triangle contained in the cordophones of the sacred instrument, it is suggested that we take into account the layout of the cordophones. The cordophones are arranged in descending order from the eighth (8th), the longest, to the shortest. Similarly, the descending order evolves from the cordophone that seems to disappear under the armpit of the sacred sculpture, to the longest. What's more, they are arranged so that a space of equal length separates them. Pascal's triangle is a *fascinating mall'K'inal object with many properties and applications"*. above:

Fig: the triangle in Blaise Pascal's treatise

Sources: https ://www. Wikipedia.org>trtangle-de-pascal

Pascal's triangle above will serve as an illustration to help us understand the nature of cordophones in sacred art. It is suggested that we consider only the triangular form initially. On the plane, the numbers evolve from zero (0) to (10), according to the vertical and horizontal orientations, so that the zero number can constitute the vertex of the triangle. Secondly, all the vertical lines must be excluded from the plane, so that we only have parallel lines, according to the arrangement of the diagonals. Note that each straight line segment is linked to the same value, both above and below, i.e. to each of its extremities. There are a total of 10 parallel line segments arranged along the diagonals of a quadrilateral. We can see that the number 10 is a smaller unit than the number 11, counting the number zero as a unit in its own right. This leads to a very simple rule: the length of a triangle within a quadrilateral represents one less of the diagonals contained within it. Quite simply, this means that the zero number at the vertex is not taken into account, suggesting that the triangle has no vertex.

We should now consider the typical example that corresponds to the arrangement of the Ngoma cordophones of the ancients. It will therefore be suggested to count from zero (0) to eight (8) on both planes, which in reality amounts to counting from 1 to 9. Given that we are in a logic of dual or binary correspondence, there will remain one unit without a binomial. Hence the numerical pairs (1;1), (2;2), (3;3), (4;4), (5;5), (6;6), (7;7), (8;8). So there are effectively 8 parallel line segments within a triangle, so the initial length is 9 perforations or 9 cm. The Ngombi cordophones symbolise the diagonals of a square.

b-) diagonals and axis of symmetry

Here, we need to complete the geometric figure in which the triangle is inscribed. To do this, we need to look again at Pascal's triangle, as illustrated above. The 10th line segment (diagonal) will be taken as the axis of symmetry: *"An axis of symmetry is a line that cuts a geometric figure into two superimposable parts. A right-angled triangle does not have an axis of symmetry. The particular triangles that have one or more axes of symmetry are the isosceles and equilateral triangles"* to complete the rest of the diagonals on the side opposite the vertex. There are exactly 9 line segments above the vertex, and a further 9 line segments below the vertex. This gives us a square with sides 11 cm long. There are 19 diagonals within this square, according to a global view exclusive to this order.

The segment of straight line passing through the centre of the base of the triangle and dividing its height in two, will be extended to form another 10th diagonal at the centre, and will in turn be taken as the axis of symmetry. In this way, there are a total of 19x4=76 diagonals in a square 11 cm long.

As far as the Sacred Harp is concerned, it is obviously the 8th cordophone that will be taken as the axis of symmetry. There are indeed 7 cordophones above it, and 7 more will be added below it, so there are 15 cordophones. However, reading the diagonals implies four possibilities, so there are a total of 15x4=60 diagonals within a 9 cm or 9 perforation square. The cordophones of sacred art are thus the product of a diagometric vision. Let's take a look at the following square:

Fig: a squared square

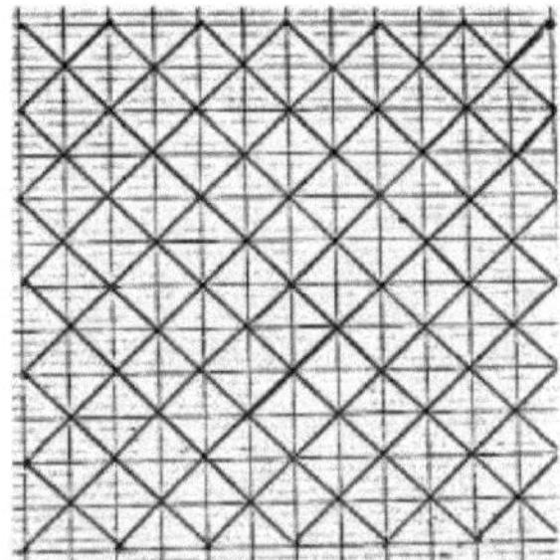

The tile above represents a complete reconstruction of the geometric figure in which the Ngombi trilateral figure was titled. The length of this tile is 7 cm, which gives rise to 6 digonal lines, according to a triangular perception. The number 6 thus symbolises the main diagonal taken as the axis of symmetry, from which we can see 5 straight segments above it and 5 below it.

The triangle in the sacred instrument is therefore really a view of half of the quadrilateral illustrated above, with particular emphasis on the main diagonal as the axis of symmetry.

c-) diagonals, axis of symmetry, triangles and fractals

Having discovered that the cordophones of the Sacred Harp are in fact a symbolic perception of the diagonals of the square, which are 9 perforations or 9 cm long, we can also see that they have three functions:

Firstly, they are the <u>diagonals of a square</u>, since they pass through the right angles and intersect at its centre while forming a right angle.

Secondly, the main <u>digonal</u> can be used to reproduce an identical image with which it is associated, on the opposite side of the image, taking it as the <u>axis of</u> symmetry.

Thirdly, the <u>digonals</u> intersect at the centre of the square, forming <u>triangles that</u> are isolated from the sides of the said quadrilateral, with their common vertex being their point of intersection, i.e. the centre of the square.

Finally, the <u>diagonal</u> as a triangular base also allows the <u>parallel</u> arrangement of other secondary diagonals to give rise to a <u>fractal</u> construction within the <u>triangle</u>. Hence the expression fractal diagometric triangle.

It is important to stress that the fractal diagometric triangle implies both notions of symmetry and parallelism, starting from the main diagonal.

5-) the square

a-) Length and diagonometric triplet

The square is obtained from the isosceles right-angled triangle with its base taken as the axis of symmetry, followed by a triplet arrangement of the diagonals in its centre.

The quadrilateral reveals that the triangle comprising 8 chords is actually extracted from the latter, the length of which is 9 perforations or centimetres (cm). The

21

eighth (8th) cord is in fact the main diagonal of the said square. There are exactly 7 secondary diagonals on either side of the main one, giving 7+8+7=22.

The number 787 is known as the ascension triplet because the number in the centre is larger than the other two numbers of the same value on either side of it. It therefore refers to a square inscribed in another square. T(n)=n+(n+1)+n equals T (7)=7+(7+1)+7=7+8+7=22. The regression triplet subtracts the middle number by the same value as the surrounding numbers. Note: T(n)=n+1+n equals T (7)=7+1+7=15.

It should be assumed that there is always a triplet buried within the tile, depending on the fragmented arrangement of the diagonals. A table of static correspondences is thus drawn up as follows:

Numerical squares and triplets

Geometric figures	Carre				
Edge length	Main diagonal	Regression triplet	Corresponding numbers	Triplets of ascent	Corresponding numbers
1	-	-	-	-	-
2	1	0-1-0	1	0-1-0	1
3	2	1-1-1	3	1-2-1	4
4	3	2-1-2	5	2-3-2	7
5	4	3-1-3	7	3-4-3	10
6	5	4-1-4	9	4-5-4	13
7	6	5-1-5	11	5-6-5	16
8	7	6-1-6	13	6-7-6	19
9	8	7-1-7	15	7-8-7	22
10	9	8-1-8	17	8-9-8	25
11	10	9-1-9	19	9-10-9	28
12	11	10-1-10	21	10-11-10	31
13	12	11-1-11	23	11-12-11	34
14	13	12-1-12	25	12-13-12	37
15	14	13-1-13	27	13-14-13	40
16	15	14-1-14	29	14-15-14	43
17	16	15-1-15	31	15-16-15	46
18	17	16-1-16	33	16-17-16	49
19	18	17-1-17	35	17-18-17	52
20	19	18-1-18	37	18-19-18	55
21	20	19-1-19	39	19-20-19	58
22	21	20-1-20	41	20-21-20	61
23	22	21-1-21	43	21-22-21	64
24	23	22-1-22	45	22-23-22	67
25	24	23-1-23	47	23-24-23	70
26	25	24-1-24	49	24-25-24	73

27	26	25-1-25	51	25-26-25	76
28	27	26-1-26	53	26-27-26	79
29	28	27-1-27	55	27-28-27	82
30	29	28-1-28	57	28-29-28	85

Source: NNANG EBANE Sosthene Tresor

This table of mathematical data proposes a practical correspondence between the length of the side of the square and the number of diagonal triplets it contains. In other words, the ability to superimpose one plane on another that is similar or identical to it is translated into arithmetical language.

Geometric shapes contain unspeakable numerical information, which is perfectly conveyed by African religious plucked instruments. Taken together, these data always define the volume of a particular element, which can reach its highest level before falling. It's true to say that life is made up of ups and downs.

In the light of the above, it is important to understand that even spoken language, strictly speaking, is the result of mathematical properties, which is why this science is most often defined as a view of the mind.

The triplets of ascension and regression, although internal to the arrangement of the cordophones, are consistent with both the increasing and decreasing evolution of the latter, just as the zither player's touch causes the strings to move upwards, but once they are left at rest, they undergo a reduced evolution.

The ratio of external and internal data allows us to understand the dynamism offered by the diagonals. Given: VE: 4x9=36 and VI: 7+8+7=22, then

VE-VI equals 36-22=14. The squared tile shows a diagonal column of 14 tiles, of which the double is 28, the triple 42 and the quadruple 56.

The number 7 is usually distributed diagonally in this sequence. We have: 7+7 = (2x7) = 14, 7+7+7+7 = (4x7) = 28, 7+7+7+7+7+7 = (6x7) = 42, and 7+7+7+7+7+7+7+7+7 = (8x7) = 56.

The maturing stage of revolution of the diagonals under a numerical perception, is concentrated in the centre of the quadrilateral. The number 8 here symbolises the original energy that gives life to the entire plan, always located at the centre of the universe. The parallel and perpendicular lines here designate all the vibrations coming from the production of this driving energy. It is an image similar to water boiling in a pan placed on a fire, or the fractal imagery produced by a pebble thrown into a stream.

The figure and its contours show that the vibration or vibratory act is a reality or a fact outside the plane. The manipulation of the cordophones by the fingers of the harp player illustrates this perfectly.

b-) Length and number of diagonals

Calculating the number of diagonals in a square, knowing the length of its sides, is made possible by a very special method of calculation, called additive calculation by degree of stagnation and numerical assimilation, note (n)stg(n). The main aim of additive calculation and subtraction by order of stagnation and numerical assimilation is to demonstrate that, despite their diversity, digits and numbers can indeed be assimilated in order to arrive at a common result. Numbers with smaller values are added together, while those with larger values are subtracted, in order to arrive at a common ideal. The additive calculation by degree of stagnation and numerical assimilation, noted (n)stg(n), is only possible within the square. The examples below are provided to help you understand the above:

The equation is :

7Stg (4)=1 2 3 4 5 6 7

= 2+(1+3)+4+(5-1)+(6-2)+(7-3)

= 2+4+4+4+4+4

= 2+(4x5)

=2+(5+5+5+5)

=2+20

7Stg(4) = 22

It reads 11 parallel lines and upper fractals (from bottom to top) plus 11 parallel lines and lower fractals (from top to bottom).

8Stg(4)=1 2 3 4 5 6 7 8

=2+(1+3)+4+(5-1)+(6-2)+(7-3)+(8-4)

=2+4+4+4+4+4+4

=2+(4x6)

=2+(6+6+6+6)

=2+24

8Stg(4)=26

It reads 13 parallel lines and upper fractals (from bottom to top) plus 13 parallel lines and lower fractals (from top to bottom).

9stg(4)=1 2 3 4 5 6 7 8 9

= 2+(1+3)+4+(5-1)+(6-2)+(7-3)+(8-4)+(9-5)

=2+4+4+4+4+4+4+4

=2+(4x7)

=2+(7+7+7+7)

=2+28

9Stg(4)=30

It reads 15 parallel lines and upper fractals (from bottom to top) plus 15 parallel lines and lower fractals (from top to bottom).

The first method of calculation above is used to find the number of parallel lines that can be drawn within squares whose sides measure 7cm, 8cm, and 9cm, etc, and which can be counted from top to bottom or bottom to top, from right to left, and from left to right. Hence the triplets 7-11-22, 8-13-26, and 9-15-30. The plane is thus divided into four equal parts by the diagonals.

The number 2 is always isolated from the rest of the calculation, so that only the binary orders of ascent and regression in a given plane are taken into account.

However, it is possible to simplify the previous method of calculation by devising an easier one. The calculation by degree of stagnation involves considering a given natural integer, from which two (2) units are subtracted, the newly obtained value is multiplied by four (4), then added by (2). The proposed mathematical formula is:

n-2=yx4=x

We have:

7-2=5x4=20+2=22

8-2=6x4=24+2=26

9-2=7x4=28+2=30

10-2=8x4=32+2=34

The table below gives an overview of this mathematical configuration:

Assessing the numbers on the board and additive and subtractive calculation

in

order of stagnation and numerical assimilation

Geometric figures	Carre				Stagnation	Double stagnation
Numbers	Internal Value 1(VI1)	Internal Value 2 (VI2)	Number of unit lines	Number of binary lines		
1	-	-	-	-	-	-
2	1	-	1	2	-	-
3	1	2	3	6	2	4
4	2	3	5	10	4	8
5	3	4	7	14	6	12
6	4	5	9	18	8	16
7	5	6	11	22	10	20
8	6	7	13	26	12	24
9	7	8	15	30	14	28
10	8	9	17	34	16	32
11	9	10	19	38	18	36
12	10	11	21	42	20	40
13	11	12	23	46	22	44
14	12	13	25	50	26	52
15	13	14	27	54	28	56
16	14	15	29	58	30	60
17	15	16	31	62	32	64
18	16	17	33	66	34	68
19	17	18	35	70	36	72
20	18	19	37	74	38	76
21	19	20	39	78	40	80
22	20	21	41	82	42	84
23	21	22	43	86	44	88

24	22	23	45	90	46	92
25	23	24	47	94	48	96
26	24	25	49	98	50	100
27	25	26	51	102	52	104
28	26	27	53	106	54	108
29	27	28	55	110	56	112
30	28	29	57	114	58	116

Source: NNANG EBANE Sosthene Tresor

If we look very carefully at this table, we can see that the square, whose sides are 9 cm long, has exactly 30 parallel diagonals. The corresponding numerical triplet is 717 or 7+1+7=15, so the number 30 also designates both the upper and lower readings, according to an ascending and descending perception of the diagonals on a plane.

Mathematics brings to the fore in this configuration the idea that the resolution of a societal problem involves the downgrading of the egos of each individual to a single sphere of thought. In other words, each individual's contribution counts towards solving the problem. A political problem should not necessarily be solved only by men from that milieu. In ancient times, the quest for knowledge led rulers to surround themselves with people of exceptional ability, who could contribute to solving society's problems for the good of all. These were not just people with an academic education, but also with inegalitarian knowledge, such as the gift of clairvoyance, the ability to read time and much more. The hierarchisation of society is not an inflexible law, when the quest for knowledge relating to the true resolution of a wide-ranging preoccupation is such a desired ideal. Everything suggests that multiplicity is merely an appearance, but that unity is the luminous focus of multiplicity. The transition from multiplicity to unity, or from unity to multiplicity, offers a visual perception from the smallest dimension to the largest, and vice versa. The cordophones of the Ngombi African spiritual instrument are effectively pinched in ascending and descending order, depending on whether the observation begins at the top or the bottom. It must be understood that a given figure or number is necessarily the result of the unity of two or more figures or numbers. A given element necessarily remains attached to a group, despite its innumerable realisations. For the universal elements all evolve within a well-defined space, and function according to a particular dynamism. Hence, self-determination is an emerging question of the pure imagination of human understanding. The universe is made up of elements that are both fragmented (different calabashes) and condensed (identical calabashes). This unity is no more and no less than the return of these same elements to the beginning, so we have to admit that the universe knows how to take care of itself; it is a machine capable of freeing itself of impurities through its functioning. It is from this perspective that mathematics is defined as a view of the mind centred around the burning question of being by being and for being, with the answer going back to the genesis of its creation. This science was undoubtedly developed not to deform the universe, but rather to understand it according to a fundamental principle of strict respect for the beings of nature. It is man's reconciliation with his ancestral origins that is expressed through this science, which relates to the unity of cordophones and geometric figures. Man remains the central object of his own sciences, a copy of the universe and also living within it. Man tries to understand himself through the sciences. We do not learn in order to succeed, but to understand ourselves, because success is not something that belongs to man, but rather to his creator. Humanity remains on a constant journey towards a common ideal: being born, growing up, growing old and dying all lead to a certain destination. Just as the conclusions of an analysis can serve as new subjects of study for other researchers, so death can be the continuation of life, which is even logical, especially as the march never ends. Humanity is subject to the cult of self-preservation in every field. Each new idea that is put forward for the good of a business is not new, because it is already designed to be known and exploited according to the reading grid specific to the fruit of patience, and not by forceps. So we need to understand that the world of intelligences exists, far beyond the human world. It is the journey to the heart of the soul that opens the doors to this mysterious destination.

c-) diagometric stagnation

When a square is crossed on both sides by straight lines parallel to its diagonals, there is a unitary and dual stagnation of the squares on the diagonals. In other words, for each number taken for the length of the sides of the tile, there is a corresponding diagonal stagnation according to a precise numerical value.

Looking at the same statistical table of mathematical data below, we should naturally consider neither more nor less than the last two columns on the right, marked stagnation and double stagnation. If we look at numbers 7, 8 and 9 as examples, we can see the following correspondence:

We have:

7-10-20-30-40

8-12-24-36-48

9-14-28-42-56

The 7 cm square is stagnating in diagonal units of 10, binary units of 20, trinitarian units of 30 and quadruple units of 40.

The 8 cm square is stagnating in the unit diagonal of 12, the binary of 24, the trinity of 36 and the quadruple of 48.

The 9 cm square is stagnating with a unit diagonal of 14, a binary diagonal of 28, a trinity diagonal of 42 and a quadruple diagonal of 56.

The table summarising the mathematical data on the tile provides an overview of the above:

The evolution of cords within the frame

Geometric figures	Carre					

Numbers	Internal Value 1(VI1)	Internal Value 2 (VI2)	Number of unit lines	Number of binary lines	Stagnation	Double stagnation
1	-	-	-	-	-	-
2	1	-	1	2	-	-
3	1	2	3	6	2	4
4	2	3	5	10	4	8
5	3	4	7	14	6	12
6	4	5	9	18	8	16
7	5	6	11	22	10	20
8	6	7	13	26	12	24
9	7	8	15	30	14	28
10	8	9	17	34	16	32
11	9	10	19	38	18	36
12	10	11	21	42	20	40
13	11	12	23	46	22	44
14	12	13	25	50	26	52
15	13	14	27	54	28	56
16	14	15	29	58	30	60
17	15	16	31	62	32	64
18	16	17	33	66	34	68
19	17	18	35	70	36	72
20	18	19	37	74	38	76
21	19	20	39	78	40	80
22	20	21	41	82	42	84
23	21	22	43	86	44	88
24	22	23	45	90	46	92
25	23	24	47	94	48	96
26	24	25	49	98	50	100
27	25	26	51	102	52	104
28	26	27	53	106	54	108
29	27	28	55	110	56	112

Sources: N'NANG EBANE Sosthene Tresor

The table above gives a practical view of each digit and number taken for the length of the side of the square, their internal values VI1 and VI2, and their stagnation in unit and binary diagonals. It offers a numerical view of the square, according to the configuration of plucked string musical instruments. Geometrical shapes are graphic expressions of arithmetical data, just as algebra is just an arithmetical expression of geometrical shapes. From this point onwards, musical instruments are therefore compilations of geometric shapes and algebraic data in the form of sculpture.

The additive calculation by degree of stagnation and numerical assimilation merely presents the result of the number of diagonals parallel to the central diagonal on either side, without however demonstrating the unity of the internal values VI1 and VI2.

However, it is obvious to observe that the diagonal stagnation of each chifife and number is revealed within this mathematical formula. We just need to exclude chifife 2 and consider only the multiplication of the values.

Example 1:

7Stg (4)=1 2 3 4 5 6 7

= 2+(1+3)+4+(5-1)+(6-2)+(7-3)

= 2+4+4+4+4+4

= 2+(4x5)

=2+(5+5+5+5)

=2+20

7Stg(4) = 22

Equivalent to :

7Stg (4)=1 2 3 4 5 6 7

= (1+3)+4+(5-1)+(6-2)+(7-3)

= 4+4+4+4+4

= (4x5)

=(5+5+5+5)

=20

7Stg(4) = 20

Example 2:

8Stg(4)=1 2 3 4 5 6 7 8

=2+(1+3)+4+(5-1)+(6-2)+(7-3)+(8-4)

=2+4+4+4+4+4+4

=2+(4x6)

=2+(6+6+6+6)

=2+24

8Stg(4)=26

Equivalent to :

8Stg (4)=1 2 3 4 5 6 7 8

= (1+3)+4+(5-1)+(6-2)+(7-3)+(8-6)

=4+4+4+4+4+4

= (4x6)

=24

8Stg(4) = 24

Example 3:

9stg(4)=1 2 3 4 5 6 7 8 9

= 2+(1+3)+4+(5-1)+(6-2)+(7-3)+(8-4)+(9-5)

=2+4+4+4+4+4+4+4

=2+(4x7)
=2+(7+7+7+7)
=2+28
9Stg(4)=30
Equivalent to :
7Stg (4)=1 2 3 4 5 6 7 8 9
= 2+(1+3)+4+(5-1)+(6-2)+(7-3)+(8-6)+(9-7)
= 4+4+4+4+4+4+4
= (4x7)
=28
9Stg(4) = 28

The triangle made up of 8 cordophones comes from a square 9 cm long whose diagonals stagnate at a binary value of 28. The numbers 8 and 7 designate the undulatory values evolving within this quadrilateral. Note that triplet 888 and triplet 789 both refer to the number 24. The first triplet is a diagonometric grip on the triangle, while the second refers to the length and the two internal waveforms. In this case, the number 888 is described as an identical or homogeneous triplet, while the number 789 is said to differ. Everything seems to indicate that the number 9, which is a supreme component of the differential triplet, has a unit removed from it in order to add it to the component of smaller value (7), in order to obtain the homogeneous or triangular and diagonometric triplet.

d-) diagonals and triangular vertices

In the light of the foregoing, the study gradually turns to the multiple function of the diagonals within the square. The diagonal arrangement of the triangle within the square is obviously accompanied by two very specific images: the triangle appears both with and without its vertex. This is particularly true of the wave values VI1 and VI2, which can be seen within the square. It is natural to propose an image perception of the above:

-triangle and flat vertex (-)

The exclusion of the diagonals within the square, which is full of many parallel line segments, leads to the emergence of triangles with flat vertices. The idea behind this geometric construction is that there is a point of ascent and regression, with no possibility of stagnation. All we need to understand is that we are analysing a trilateral figure, which is also accompanied by a simple triplet logic: ascent, stagnation, regression. Now let's look at the next square:

Fig: the square with 9 perforations/cm

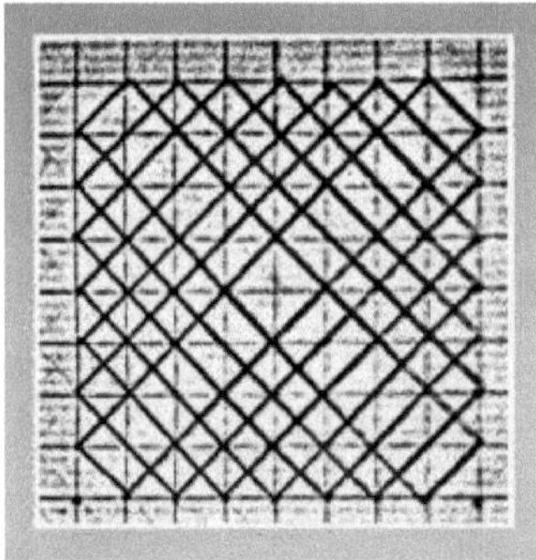

The 9 cm square above is used as an illustration. If we exclude the diagonals relating to the number 8, we can see that there are only 7 straight line segments left, arranged in four similar planes. The seventh (7) diagonal now forms the base of the various triangles with flat vertices. The result is four (4) triangular triplets of the number 777.

We can see that the sides of the triangle are the same length 7 against 7, but that the base is double 14. This is therefore an isosceles triangle. Let b(base)=c(sides) x c(sides).

The triangle shown above is just a linear numerical correspondence, without a value other than 0 occupying its vertex. From the above, we can deduce that, depending on the configuration of the triangular digits, the figure has a flat or hollow vertex. In view of this particular feature, the middle of the triangle can be used to irrigate rainwater, or to channel solar energy in order to capture it for scientific purposes.

It is quite possible that human anatomy may also find a particular echo in this geometric figure, as a symbol of the female sex capable of receiving the man's semen in order to give birth to a child later on. This type of triangle would correspond perfectly to the mortar in African culture, in which food is processed, enabling it to be transformed from its fragmented state into a condensed one.

triangle with pointed top $( )^{\pi}$

It has been shown above that the number symbol of the main diagonal is less than one unit shorter than the unit length of the square. The dance seems to continue with the triangle appearing with and without a vertex. Let's take a look at the square above:

Fig: a squared square

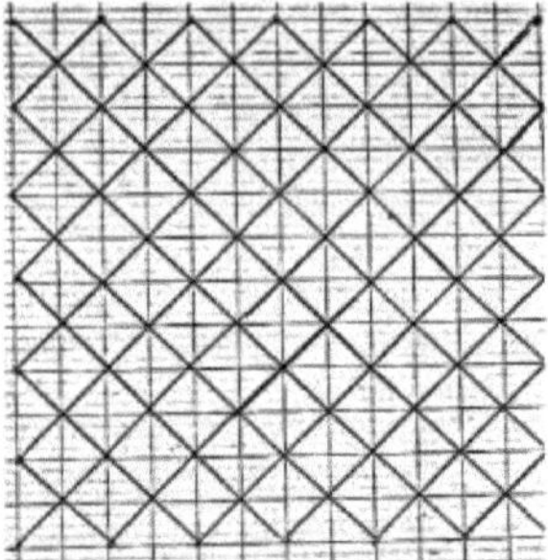

The quadrilateral shown here is 7 cm long. However, the triangular socket reveals two sides which are 7 cm long, but whose base measures 13 cm. You can see that the number 7 is the centre of the axis of the base, and therefore the number 13.

We can see that the presence of diagonals within a square is accompanied by triangular vertices, while determining the length of the side as the symmetrical centre of the middle of its base. The base of the triangle contained in this 7cm long square has the numerical triplet 7+1+7=13.

Once a number has been added to the vertex of the triangle, the lengths of the sides increase. The vertex now has a point, which differentiates it from the first. With this in mind, the triangle can be used as a symbol of Masculinity. In this way, it can be compared to the pestle used to crush food in a mortar.

The two types of triangle therefore describe a single element capable of assuming a contended shape for a moment, before assuming a flat shape the next. Isn't it said that anything that can stand upright always ends up collapsing? However, we're dealing here with the usual action of daybreak and nightfall. It's a unique sky with so many variations in light. The geometric shapes give off imagery that can be transposed to everyday life. However, both the triangles and the cordophones fulfil the same function of ascent and regression.

e-) diagonals and rectangle

Extracting the diagonals within the square reveals the presence of another quadrilateral within it. We discover that it is not just another square that is inscribed within the square through the triplet perception of the diagonals, even less so the trilateral figures, but also the rectangles. To achieve this, we suggest examining the square below:

Fig: the square with 9 perforations/cm

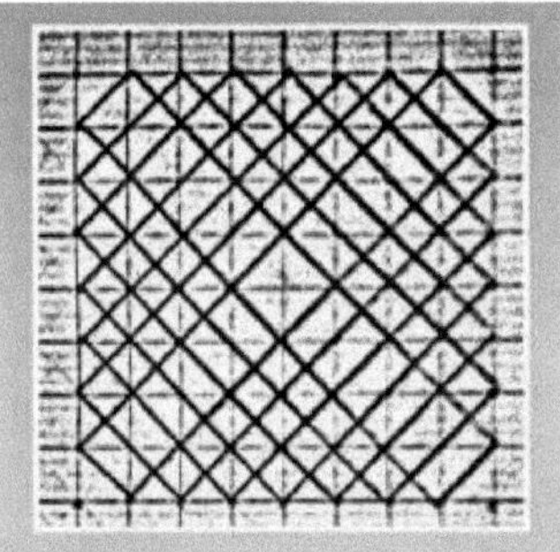

Excluding the diagonals results in the formation of rectangles surrounding the central square on either side. There are exactly twelve (12) rectangles in rows of six (6) blocks in the geometric figure above. The linear diagonals have been removed, but the square and its rectangles still follow their shape. We can see once again that the rectangle is a summation of two squares, by extracting the diagonals from the figure under examination.

Triplicity is not only relative to the three (3) sides of the triangle, but also to the unity of the rectangle with the square, referring to a triplet perception of the second quadrilateral. According to a general perception, the square, the rectangle and the triangle are all inscribed within a basic square. In this way, the number five (5) is revealed as a numerical symbol for the compilation of geometric figures.

The multifunctional nature of diagonals is quite simply impressive: the digonals of a square, the base of a triangle, the axis of symmetry, the equal sides of an isosceles triangle, parallel straight lines, perpendicular straight lines, and so on.

F-) square and triangle values

It has been observed that the square and the triangle make up the majority of the pyramid, so it occurred to us to want to find out the numerical value of each. By calculating the number of squares each triangle comprises within the square. We need to understand that the triangles rise above the quadrilateral to give it height, but that each triangle has the same number of squares. To achieve this, consider the length of the side of the square, subtracting first one unit, then two units. These different values are multiplied together and the whole is multiplied by two. NC=[(Vn-1) x (Vn-2)]x2.

Consider the tile below. The first column of tiles in the vertical plane is twice as small as the length of the side of the tile, so VI1=Vn-2. On the other hand, the second column contains a number of tiles that is one less than the length of the tile. VI2=Vn-1. The length of the tile is 7 points, let's say 7cm, the first column contains 5 tiles, the second contains 6 tiles. The total number of squares is obtained using the formula: NC=[(Vn-1)x(Vn-2)] x2 equals [(7-2)x(7- 1)]x2=(5x6)x2=30x2=60

Fig: a squared square

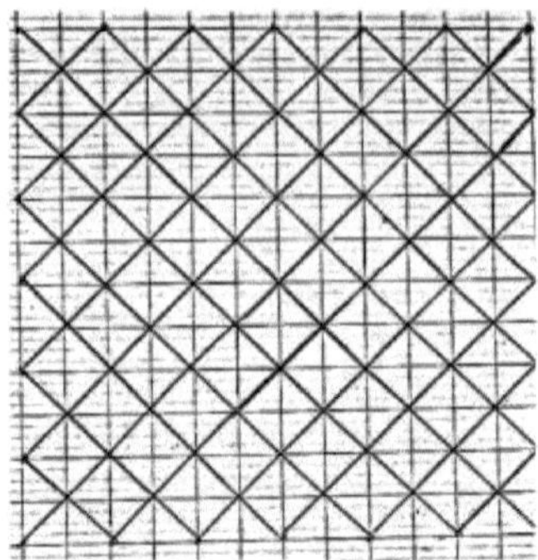

Source: N'NANG EBANE Sosthene Tresor

If you look very carefully at the tile above, you can see that the diagonals intersect in the middle, forming four triangles on each side. To determine the number of tiles in each triangle, divide the result of the formula NC=[(Vn-1)x(Vn-2)]^x 2 equals [(7-2)^x (7- 1)]^x 2=(5^x 6) 2=30^{xx} 2=60, by 4. Then T=NC/4=60/4=15. If the tile has 7 points as the length of the sides, then the triangles are each made up of 15 tiles.

For 8 as the edge length, we have: T (7 6) 2 42 2 82:4 21

For 9 as the edge length, we have: T (8 7) 2 56 2 112:4 28

For 10 as the edge length, we have: T (8 9) 2 72 2 144:4 36

The table below illustrates this perfectly:

The evolution of cords within the frame

Geometric figures	Carre					Triangle
Numbers	Internal Value 1 (VI1) n-2	Internal value 2 (VI2) n-1	VI1+VI2	VI1XVI2	Multiplication by 2	Divion par 4
1	-	-			-	-
2	0	1	1	1	2	0,5
3	1	2	3	2	4	1
4	2	3	5	6	12	3
5	3	4	7	12	24	6
6	4	5	9	20	40	10
7	5	6	11	30	60	15
8	6	7	13	42	84	21
9	7	8	15	56	112	28
10	8	9	17	72	142	35,5
11	9	10	19	90	180	45
12	10	11	21	110	220	55
13	11	12	23	132	264	66
14	12	13	25	156	312	78
15	13	14	27	182	364	91

16	14	15	29	210	420	105
17	15	16	31	240	480	120
18	16	17	33	272	544	136
19	17	18	35	306	612	153
20	18	19	37	342	684	172
21	19	20	39	380	760	190
22	20	21	41	420	840	210
23	21	22	43	462	924	231
24	22	23	45	506	1012	253
25	23	24	47	552	1104	276
26	24	25	49	600	1200	300
27	25	26	51	650	1300	325
28	26	27	53	702	1404	351
29	27	28	55	756	1512	378

Sources: N'NANG EBANE Sosthene Tresor

The table shows the transition from square to triangle according to specific values for the number of squares. The column of data multiplied by 2 refers to the total number of squares that make up the triangles of a square of the corresponding length. The column of data just below the division by 4 indicates the exact value of a single triangle.

g-) Ngoma tripartite configuration

By always basing the analysis on the evolution of the diagonals within the square, we should now no longer be interested in the ascent and regression taken together, but rather in one part of the plane. By taking the diagonal as an axis of symmetry, the isosceles right-angled triangle has led to a square. We need to adapt the additive calculation by degree of stagnation and numerical assimilation to the triangle, according to the configuration of the Sacred Harp triangle comprising: the upper perforations, the strings in the centre, and the lower perforations.

The corresponding mathematical formula is simplified as follows: a given digit or whole natural number is subtracted by one unit (1), the new value obtained is multiplied by three (3), then its tripartite distribution over the entire half-plane is obtained.

We pose:

7-1=6x3=18

(6 people sup-6 cordophones-6 people inf)

8-1=7x3=21

(7 sup-7 cordophones-7 inf)

9-1=8x3=24

(8 people sup-8 cordophones-8 people inf)

10-1=9x3=27

(9 sup-9 cordophones-9 inf)

If we consider the different results obtained, we can see that the Ngombi with 8 cordophones is extracted from the square with 9 cm of sides, because it has 8 upper perforations, 8 cordophones and 8 lower perforations, i.e. 3x8=24.

However, it should be remembered that the number of vices used to fine-tune cordophones is exactly the same as the number of perforations and cords. With this in mind, we just need to replace the number 3 with the number 4.

We'll have:

7-1=6x4=24

(6 upper peris-6 cordophones-6 lower peris and 6 vice)

8-1=7x4=28

(7 over-7 cordophones-7 under and 7 vice)

9-1=8x4=32

(8 sup-8 cordophones-8 inf and 8 vices)

10-1=9x4=36

(9 sup-9 cordophones-9 inf and 9 vices)

In view of the above, the triangle and the square are combined through the cordophones, perforations and vices of the Ngombi.

By extension, the additive calculation by degree of stagnation and numerical assimilation is effectively consistent with the above. All we need to do is make a few changes to the condensations to bring out the table of multiplication by 4.

We pose:

7stg(4)=1 2 3 4 5 6 7

= (1 +3)+(2+2)+(3+1)+4+(5-1)+(6-2)+(7-3)

=4+4+4+4+4+4+4

=4x7

=28

7Stg(4)=28

8stg(4)=1 2 3 4 5 6 7 8

= (1 +3)+(2+2)+(3+1)+4+(5 -1)+(6-2)+(7-3)+(8-4)

=4+4+4+4+4+4+4+4

=4x8

=32

8Stg(4)=32

9stg(4)=1 2 3 4 5 6 7 8 9

= (1+3)+(2+2)+(3+1)+4+(5-1)+(6-2)+(7-3)+(8-4)+(9-5)

=4+4+4+4+4+4+4+4+4

=4x9

9Stg(4)=36

It should be borne in mind that the results of this method of calculation vary according to the configuration in which its components are organised. Mathematics is not an exact science in the strict sense of the term, because it depends on particular configurations. It seems that it was developed to understand how time works, so the unpredictable nature of nature means that we obtain results that are not always the same.

In reality, the multiplication tables allow us to stagnate or fix on a specific number, but beyond that, the figures that follow are only negative. Everything is reminiscent of the transition from the vertex of a triangle to its base. They suggest a uniform perception of the elements of the universe. Everything happens as if the stagnant number occupies the value 0 in relation to its stagnation vector.

Example 1:

7Stg(6)=1 2 3 4 5 6 7

= (1+5)+(2+4)+(3+3)+(4+2)+(5+1)+6+(7-1)

= 6+6+6+6+6+6+6

= 6x7

=42

7Stg(6) = 42

7Stg (5)=1 2 3 4 5 6 7

= (1 +4)+(2+3)+(3+2)+(4+1)+5+(6-1)+(7-2)

= 5+5+5+5+5+5+5

= 5x7

=35

7Stg(5) = 35

Example 2:

8Stg (7)=1 2 3 4 5 6 7 8

= (1 +6)+(2+5)+(3+4)+(4+3)+(5+2)+(6+1)+7+(8-1)

= 7+7+7+7+7+7+7+7

= (7x8)

=56

8Stg(7) = 56

8Stg (6)=1 2 3 4 5 6 7+8

= (1 +5)+(2+4)+(3+2)+(4+2)+(5+1)+6+(7-1)+(8-2)

= 6+6+6+6+6+6+6+6

= (6x8)

=48

8Stg(6) = 48

Example 3:

9Stg (8)=1 2 3 4 5 6 7 8 9

= (1+7)+(2+6)+(3+5)+(4+4)+(5+3)+(6+2)+(7+1)+8+(9-1)

= 8+8+8+8+8+8+8+8+8

= (8x9)

=72

9Stg(8) = 72

Or

9Stg (7)=1 2 3 4 5 6 7 8 9

= (1+6)+(2+5)+(3+4)+(4+3)+(5+2)+(6+1)+(7-0)+(8-1)+(9-2)

= 7+7+7+7+7+7+7+7+7

= (7x9)

=63

9Stg(7) = 63

Examination of the digital condenser

Example 1

1 2 3 4 5 6 7 (stagnation vector 7)

5 4 3 2 1 0 -1 (decrease in -1)

6 6 6 6 6 6 6 (stagnating)

Example 2

1 2 3 4 5 6 7 8 (stagnation vector 8)

4 3 2 1 0 -1 -2 -3 (decrease in -3)

5 5 5 5 5 5 5 5 (stagnant figure 5)

Example 3:

1 2 3 4 5 6 7 8 9 (stagnation vector 9)

6 5 4 3 2 1 0 -1-2 (decrease in -2)

7 7 7 7 7 7 7 7 7 (Stagnant number 7)

The different lines of calculation show that there are three types of evolution: growth, regression and stagnation. However, these three (3) types of evolution are present within the triangle and the square. The introduction of cordophones into geometric shapes reveals existential lessons. The diagonals parallel to the main one represent both regression and ascent, while the main diagonal defines stagnation or maturation. It is a stem energy that sends out its light beams on either side of where it radiates. In simple terms, humans have drawn their knowledge from an intelligence older than their appearance on the world.

h-) pyramidal and fractal vision of Ngombi

After a lengthy presentation of the different mathematical contours of this ancestral sacred art, it is time to present its pyramidal and fractal view. Knowing that the length of the square is 9 perforations, we will see that 8 squares in fractal posture evolve within it, excluding the diagonals once again. The following is a clear illustration of the above:

Fig: the fractal pyramid

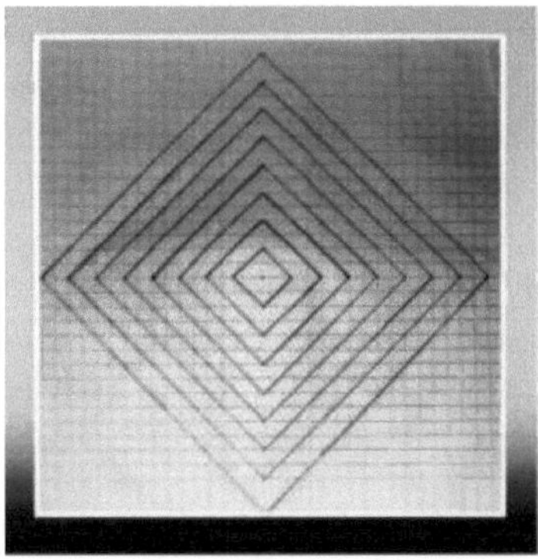

The resulting image is an absolutely sublime pyramid. The geometric figure is thus seen from the heavens, and actually has 8 fractal squares. This is how a supposedly complete view of the traditional Ngombi instrument is obtained. The observation of the mathematical view of the Sacred Harp thus concludes at the heart of the pyramid. Hence, the civilisational memory relates to a pyramid equipped with a sound system. So what's the point of associating geometric form with echo? Can we not see in it the din emanating from the jet of the Nile in the heart of the Mediterranean?

i-) image of the Sacred Harp on Egyptian frescoes

Having focused on sacred art through a strictly mathematical analysis, leading to the discovery of the pyramid as a complete geometric figure based on its triangular perception, we now turn to its practical image in Egyptian frescoes. The photograph below illustrates without a shadow of a doubt what has just been said:

Source: unknown

The photograph shows the hieroglyphs and the Sacred Harp held by an individual in the centre, between two other figures, one sitting in front of him and the other just behind. The arrangement of the individuals in the frescoes always relates to the notion of triplicity, which seems to have been considered a highly sacred posture. Sacred art relates not only to the expression of spirituality according to mathematical codes, but also to the incorporation into it of the region of origin of its owner people. It is in fact from this perspective that art is defined as the writing of the course of a particular history captured by the eye of the observer on a durable, papal and easily transportable medium, in order to communicate it to future generations. The incorporation of the echo into the solid form that is the pyramid can only relate to the invisible giving rise to the physical object by means of vibration, of which the jet of the Nile in the Mediterranean and the pyramid are practical expressions of theological thought. The unconditional preservation of a people's memory can be summed up in the art of survival. From this point of view, the sacred instrument seems to be defined as a mark of identity, but what about the Mvet Ekang?

II-) MVET EKANG

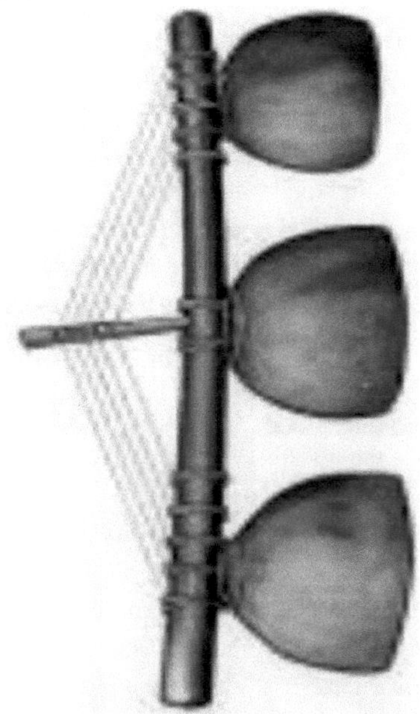

https://www.Fr.m.wikipedia.org/wiki/Mvett-Ekang

our late

gabonese

politician, man of culture

and musician,

ZENG EBOME PIERRE CLAVER

(The Fang philosopher)

((19/09/1953)-(19/05/2022))

1-) expression of the quantity

a-) the qualifier Mvet

Mvet Ekang as a qualifier or noun refers to the expression of greatness, but one that requires a great deal of personal effort. It is a vision close to the dreams that we wish to realise as human beings, while being aware of the fact that we will have to move heaven and earth to get there. The African theological vision, knowing that man's intentions are veined without recourse to the spirit world, from which the destruction of matter is imposed:

"Mvett is said to come from the verb "a vet", which means "to stretch" or "to s^lever". It is primarily about an individual or collective spiritual ëK'yaⲠon."

At first glance, the verb "to stretch out" conveys the image of a person in a posture that gives him or her a greater height than that described as ordinary. It implies two respective boundary lines, the starting line and the finishing line. Humanity naturally stands upright at a certain height, so stretching gives it a supreme height. We are therefore talking about a single individual capable of experiencing two evolutionary periods. The ascent will be taken as the supreme height, while the regression is relative to his ordinary height. The Mvet therefore refers exclusively to the expression of height.

The second view focuses on the manifest knowledge of humanity, as made up of body and spirit, while highlighting the fact that it is the product of the invisible. This superior world must be visited in order to better dominate the matter in which our spirit evolves. Literally speaking, Mvet encourages the spirit to escape from the body from time to time in order to reconnect with its essential universe. This is a more than obvious vision of the human body as a transport vehicle for the spirit, the first having a limited course, while the other sets off on an infinite race. A single element undergoing two opposing evolutions. Humanity is in effect a divinity that must always connect with the invisible universe, so as not to lose its status as such.

b) morphology

Civilisational memory relates not only to the idea of grandeur in the etymology of the name, but also to its morphology. Indeed, the instrument reveals a mountain backbone through the triangle that makes it up. The cordophones clamped to it offer the possibility of both ascent and regression. It's worth pointing out that although the mountain is a terrestrial element, its verticality unites the earth with the sky. What's more, reaching its summit provides a sovereign view over the entire surface of the earth, which is how the subject expresses the idea of supremacy. The eye, as an organ of the human body, is ultimately revealed as a window to memory. Remembrance is naturally and rightly the final word in the story, brought to the fore by this ancestral jewel. The memory of those who conquered it, the memory of the greatness for which it is intended, the memory of life after death, the memory of its homeland, and more. The image of the mountain used to understand the Mvett calls on our knowledge of the elements of nature that can be used as symbols of power, through memory. Memory, or recollection, remains frozen when its existence is unknown, but knowledge of it leads the explorer to evolve with the times as he discovers them. The encounter with a memory is a story that never leaves us, insofar as we can always enter and leave it at the same time, without ever remaining the same. Mvet is a permanent exorcism of the superfluous clumsiness of human thought.

c-) verticality of components

The components of the ancestral jewel are all related to the elements of nature, more specifically the forest. The forest is a toolbox for the Fang people in particular, and for all other peoples in general. The memory evoked above imposes itself de facto on the unconditional respect for this living environment, and its infused knowledge, as witnessed by the components of sacred art:

"The instrument has six main parts: 'a dry bamboo rod'; 'sound strings'; 'rattan-

wire rings'; 'calabashes', and 'a lanyard' which is attachedë both ends to the bamboo rod and allows the instrument to be easily carried as a sling."

It is important to stress in passing that one of the characteristics common to all forests is the verticality of the trees. This reconciles the practical parallelism existing between the sky on the one hand and the earth on the other, giving rise to two parallel perpendicular straight lines, with another known as vertical. In the same vein, the constituents are all relative to this schematic view of the unity of opposites, as they belong to plant species tending towards the heavens. Verticality expresses life based on an exchange of opposing energies, under the permanent adaptability of the latter to a vital environment destined for them. The transformation of matter into spirit or of spirit into matter takes place vertically. The sun's zenith position can indeed be seen as a vertical opening from heaven to earth, according to the vision of a certain very advanced science. The idea that can be added to this is that mankind remains on a constant journey along a path that was mapped out long before its existence, through its many dimensions. Verticality in itself evokes the guidance of the human spirit towards the high heavens, under the guidance of an intoxicating spiritual force which it cannot but indulge. It is the unconditional return of the human soul to its universe of essence.

2-) The nature of the trilateral figure

a-) the equilateral triangle

The observation made about sacred art is based primarily on the trilateral figure, which is one of its most striking features at first glance. A closer look at the sacred art on the cover reveals that the cordophones are clamped onto the horizontal branch and at the same time pass through the easel, forming a triangle. To determine the nature of the trilateral figure, it is worth pointing out that the perforations on either side of the bridge (into which the cordophones are clamped) and those on the bridge are exactly the same number. In simple terms, if we count 5 perforations to the left of the bridge, we count exactly 5 perforations to the right of the bridge and another 5 perforations on the bridge itself. From the above, we can deduce that this is an equilateral triangle:

"Among the trilatëres figures, the ëquilatëral triangle is the one that has its three côtës ëðaux"

Of course, it could be argued that the 5 perforations in the bridge that make up the height of the triangle relate to the size of its base. On the contrary, the 5 against 5 lower perforations on the horizontal leg would simply designate the lengths of the opposite sides in height. So the arithmetical triplet 555 would be taken as the symbol of the equilateral triangle, with each number 5 referring to the length of each of its sides. From this point of view, the trestle would only serve to give height to the cordophones, a simple triangular shape.

b-) the isocele triangle

The triangular isocele obviously gives the bridge a very special configuration, symbolising its height. This particularity relates to the position of the bridge in

43

height and in relation to its base, i.e. its central position on the horizontal branch. In fact, if we consider the bridge at the centre as standing on another perforation, we would have 5+1+5=11 perforations at its base, and 6 perforations on itself, so the total number is 5+(5+1)+5=5+6+5=16. The length of the base would therefore be 16 perforations or cm, while the length of each of the sides would be 5 perforations or cm each. Hence, from this point of view, it is an isosceles triangle:

"Among the trilateral figures, the isosceles triangle is the one with two equal sides.

In view of the above, triplet 565 would refer to an isosceles triangle, according to the configuration of the Mvet Ekang instrument. It is important to stress in passing that the triangle contained in the civilisational memory contains two types of numerical triplet, called ascending and descending. The descending triplet is noted: T(n)=n+1+n equals 5+1+5=11. On the other hand, the ascending triplet is noted: T(n)=n+(n+1)+n equals 5+(5+1)+5=5+6+5=16. The first only refers to the presence of a "right angle" with no possibility of ascent, whereas the second clearly accounts for the latter.

c-) the height of the triangle

The notion of triplicity, according to the observation made of the civilisational memory, is revealed as being inscribed in the object of study in an artisanal manner. In fact, if we consider the easel as an object independent of all the other components of the Mvet instrument, we see that it is a single element with openings allowing a view of the other side. In other words, looking through the object results in an identical view, whether you are on one side or the other, in relation to the object. The view from the instrument seems to be channelled towards a precise objective. From this point on, initiation rites are a practical example of a vision of the spirit world that differs from that of ordinary mortals. It should be noted that this vision excludes even matter, which is nevertheless the bearer of the opening or perforation. Triplicity therefore brings to the fore the idea of the destruction of matter, through a view of the mind.

The triangle as a geometric figure is simply a mathematical translation of this theological thought. Its three sides are in fact united, while forming a mountainous structure with a hollow within it. This void is not a void, but rather a question that leads to a certain answer, that of the longevity of the human soul beyond the vile desires of the body. The invisible world, or the so-called spirit world, only recognises what is similar to it within man, while excluding the body, which is not an integral part of its environment. The triangle thus emerges as a reminder to humanity of the place to which it belongs and its place of essence.

3-) the order of the calabases and the triangle

a-) the central calabash

Everyone agrees that the Mvet Ekang instrument is triangular in shape, according to the arrangement of the cordophones, the bridge and the horizontal branch. However, no one has any idea about the order of the calabashes. With this in mind, it is worth looking at the instrument, which is made up of a single central calabash.

This is strictly linked to the bridge in the centre, so that it forms an extension of the bridge at the lower level. The whole plan of sacred art is related from this very moment to a trilateral figure, whose height is very much extended below its base, so that by unifying each extremity of the base at this height with straight line segments, we obtain a square. The photograph of the civilisational memory below illustrates this perfectly:

Photo 1: The Mvet has half a calabasis

https://www.Fr. m. wikipedia.org/wiki/Mvett-Ekang

Numerically speaking, this combination of trilateral figures would refer to the Purification of two glued 4s. If we consider the number 44, and change the position of the second number 4 so that it points to the right, we will obtain the image of the triangle described here.

Many people agree that the number 3 symbolises the triangle, while the number 4 denotes the square. However, the former offers only a symbolic perception in terms of its sides and angles, while the latter offers a practical view of half the triangle, from a morphological point of view. So the number 4444, made superimposable and unified in the form described above, i.e. 44 above against 44 below, refers to the unity of four triangles joined by their vertices at the centre of the square, the most reductive form of which is the square and its diagonals.

The Ngoma, which is the first subject mentioned, rightly comprises 8 upper perforations, 8 cordophones, and 8 lower perforations according to a first reading. If we include the number of vices, there are now 8 vices, 8 upper perforations, 8 cordophones and 8 lower perforations. The two sets reveal an evolution from the triangle with the numeric symbol 888 to the square with the arithmetic symbol 8888. In the same vein, the number 4444 denotes the unity of the square and the triangles, while the number 444 refers to the practical unity of three (3) triangles. Furthermore, the number 888 is double the number 444, and the same is true for the numbers 8888 and 4444, hence the Ngombi of the ancestors is truly a spiritual reality overflowing with the practical, geometrical and arithmetical structure of the pyramid. Pharaonic Egypt is even more pronounced as its universe essence.

The number 444 can also sum up the pyramid on its own, through the unity of the square and four triangles. In fact, the homogeneous triplet relates to the number 12 as both product and sum. The newly obtained number relates to the sides of each triangle unified with the square and to the sides of the square itself. However, it is important to note that in this configuration, the base of each of the triangles is unified at each side of the square. The base of the triangle is equal to the side of the tile, or the side of the tile is the same as the side of the triangle.

The Sacred Harp and the Zither Harp, traditionally called Ngoma and Mvet Ekang respectively, share a similar arithmetical configuration at this point in the analysis. The triplet 555, referring to the left, centre and right perforations, obviously symbolises the triangle. However, if we add the cordophones, which are exactly the same number, we obtain the number 5555. In view of the above, the number 555 and the number 5555 both define the arithmetical unity of the triangle with the square, hence the pyramid. The Harp Zither bridge relates to the Sacred Harp neck, or the Sacred Harp neck is synonymous with the Harp Zither bridge.

b-) the large central calabash and the other two

We now need to understand the arrangement of three calabashes, depending on whether the one in the middle is larger than the other two accompanying it, one to its right and the other to its left, both being of the same size. In fact, the simple fact of arranging three round elements in a row, depending on whether the one in the centre is larger, tends to reveal an image of it as being further forward than the other two. The size of the elements, from their overall alignment towards a solitary race forward, gives the impression of an element wanting to extract itself from the group. A visual perception of what has gone before is an obvious fact:

Photo 2: The Ekang Mvet of three calabashes

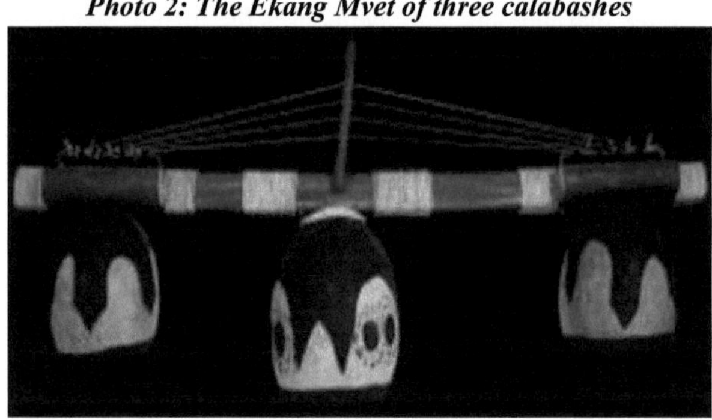

https ://www.fr. m.wikipedia. org>wiki>mvet/(2018)

According to this photograph, the central calabash is obviously more prominent than the other two, although they are all connected to the horizontal branch. In view of the above, the shape is triangular. A geometric perception of this configuration of calabashes can be illustrated as follows:

Fig: a triangle with a pointed base

46

Some might argue that this is a square and not a triangle, which is understandable, but only in part. However, in keeping with our logic, we can see that the horizontal branch is not present on the geometric figure, as it symbolises its diagonal. By inserting it into the plan, the configuration of the calabashes becomes more explicit. The main calabash designates the most prominent right angle, while the other two relate to the right angles in the background. So, the traditional Mvet Oyeng instrument, through the alignment of the calabashes seen in the photograph, marks half of a square running from its horizontal diagonal, joining the two right angles, towards the top of the third angle. The horizontal branch to which the half-calabashes are attached is a vertical view of the horizontal diagonal of the square, in this configuration.

c-) identical gourds

To conclude our analysis of the alignment of the calabashes, we need to consider their homogeneity. In fact, half-calabashes in sacred art sometimes all appear to have the same proportion. It is right to reconsider the half calabashes described above and replace the main one with another of equal size to the other two, or simply have it adjusted to the same alignment as the other two. This will obviously lead to the alignment of the gourds being confused with the horizontal branch to which they are closely linked. A visual perception of the above is of the utmost importance:

Photograph 3: The Mvet Ekang has three calabashes

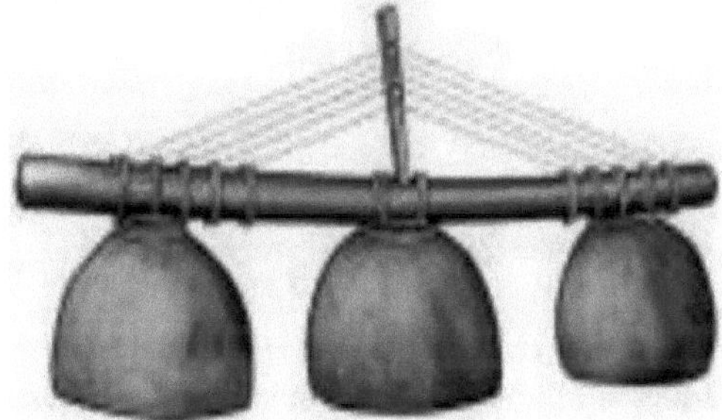

https://www.Fr. m. wikipedia.org/wikiMvett-Ekang

This photograph shows that the half calabashes are well aligned with the horizontal branch, resulting in a geometric image that is the opposite of the previous one. The digaobale and the calabashes will all be confused. The following is a perfect illustration of this analysis:

Fig: a triangle with a flat base

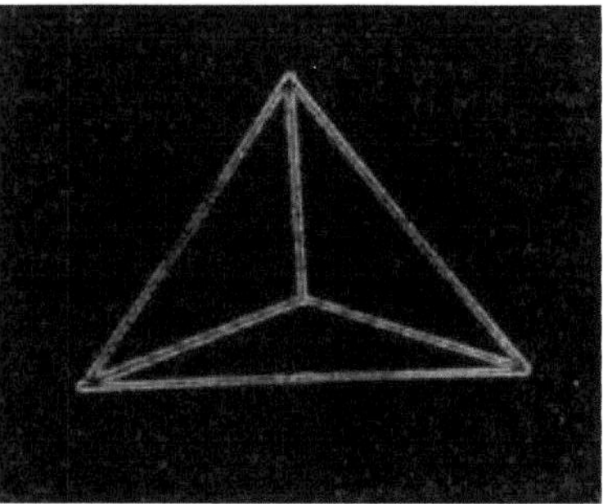

Source: NNANG EBANE Sosthene Tresor

There is not the slightest doubt that the forward angle is excluded from the geometric figure. It is easy to understand that the identical calabashes would even refer to a Mvet Ekang instrument that had no calabashes. Civilisation's memory shows that three aligned elements of the same portion reveal the image of a straight line. It is not at all appropriate to set up a certain representation, without knowing the idea that we obviously wish to express. So, the flat line can describe the absence of disturbance, while the broken line confirms the presence of a certain amount of mobility.

d-) wave variations and the pyramid

It should be pointed out that the most prominent angle and its return to homogeneity with the other two, relate to what we have called wave variations. This up and down characteristic is always linked to the square that forms the base of the pyramid. But what is the idea being expressed? The main aim of civilisational memory is to remind people of their region of origin, not only from the point of view of migration history, but also in a geometric language. So, the undulatory movements combined with the quadrilateral as the base of the pyramid relate to the idea that the pyramid was born from the Nile. The pyramid is therefore a product of the great Nile. This theological idea that water is the creative material of all things is translated into mathematical codes. The cordophones symbolise the mathematical and artistic perception of the river's waters. The waters of the Nile must always flow and be celebrated, in order to give life to the pyramid, because the solid is always born from the liquid. The multiple melodies produced by the touch of skilful fingers imply that the vital energy that makes the waters undulate is not the work of a mortal. The symbol on the cover page of the work should be taken as a starting point:

Fig: Dz^ Ayem

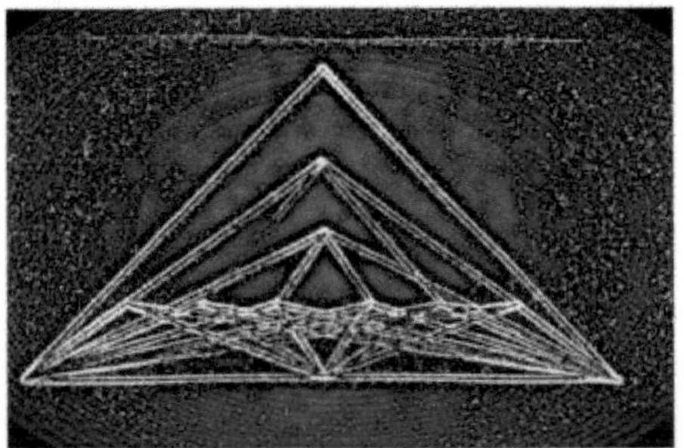

The 'Dz^ Ayem' symbol combines three important elements: the triangle, the water and the sky. The triangle alone is considered to be the whole of the pyramid, even though it is only half of it. It is therefore taken to be a solid. We can see that the water inside the pyramid appears in an undulating form, occupying its entire base. All this suggests that the pyramid seems to be gradually withdrawing from the animated water. The vibration is linked to the very work of creation, as evidenced by the undulating surface of the water on the symbol. The eye symbolises the witness to the origin of creation, the one that even refers to the notion of remembrance. The Ekang Mvet emphasises this remembrance from both an oratorical and a morphological point of view. The pyramid symbolises an eye that is wide open to the world, a testimony to the high spiritual concentration in this part of the world at the time. However, the departure of the many peoples from this part of the world towards other denominations is exclusively due to the displacement of this spiritual concentration towards the current region of Central Africa, of which Gabon is an exponentially luminous territory.

4-) Inverse triangles

a-) triangular base and axis of symmetry

Looking at the arrangement of the cordophones within the bridge and the horizontal branch, it is clear that the Mvet instrument is triangular in shape. It has been shown that the instrument on the front cover has 5 perforations on the left and 5 perforations on the right of the bridge, which itself has 5 other perforations. Hence, the triplet 555 symbolises the trilateral figure. The above can of course be illustrated as follows:

A right angle

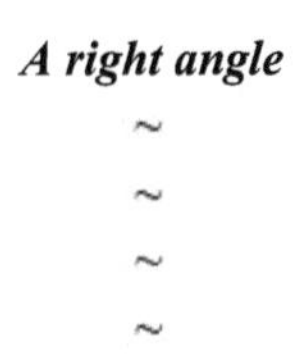

Source: NNANG EBANE Sosthene Tresor

49

The geometric shape of the right angle can be seen by counting the number of perforations in the sacred instrument. If we consider the horizontal branch with 5 plus 5 perforations in total as the axis of symmetry, because it forms the base of the triangle, we obtain exactly the same trilateral figure at the lower level. It is important to note, however, that the calabashes are not used in this configuration. There are now 5 perforations to the left and right of the bridge, and 5 perforations on the bridge itself at the top level, and 5 perforations to the left and right of the bridge, and 5 more on the bridge itself at the bottom level, where the square is revealed. The edge is therefore noted in fractional form, i.e. 555/555=1. The following image is thus obtained:

the diagonals

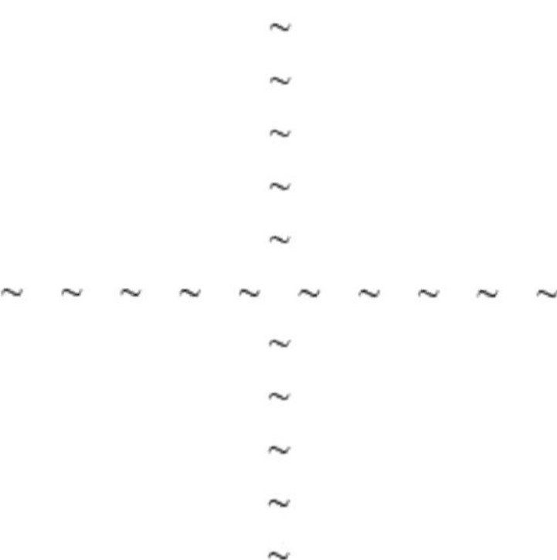

Source: NNANG EBANE Sosthene Tresor

Remembering the number 4444 as the numerical value of the pyramid, the position of the diagonals on the plane shows once again that the number 4 can be inscribed 4 times, according to both the upper and lower planes. The morphology of the Sacred Mvet Ekang, in the light of the above, brings to the fore the idea that the right angle is a diminutive vision of the diagonals. On the contrary, the latter are a duplicate of the former. If the right angle relates to unity, and the diagonals define duality, then the two forms taken together relate to triplicity. The right angle and the diagonals form a unified whole, demonstrating that the elements of nature are present in both condensed and fragmented forms. In a higher dimension, this fact suggests that nature knows how to take great care of itself, without any outside help.

b-) cordophones and diagonals

The general function of the cordophones is to reconcile each of the three perforations in a gradual order. So, by joining the different points of the plane by straight line segments symbolised by the cordophones, we obtain a fractal tile. In other words, you can watch the tiles grow and decay at the same time. From this moment on, the sacred instrument reveals itself as the expression of universal reality, unifying the two orders of revolution into one. The following is a perfect illustration of this idea:

Fig: the fractal square

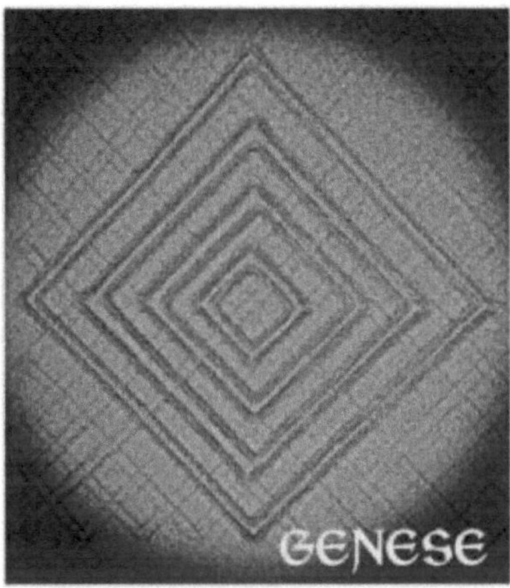

Source: NNANG EBANE Sosthene Tresor

The tile shown here represents a Mvet instrument made up of four cordophones. A single square can be seen growing and shrinking in several dimensions. This explains the notion of fractal geometry. In the light of the above, we can see that the Mvet Ekang is half of a pyramid, which can only be completed by excluding the calabashes, but also and above all by taking into account the alignment of the perforations. The musical pyramid is the second name for the sacred instrument. We have already made the link above between the echo and the physical form, through the vivifying and the vivifier, to understand civilisational memory from this angle.

c-) triangles and the balance of forces

It is quite clear that the previous quadrilateral is a square. However, it is noticeable that it occupies a vertical posture, which is not the one we are usually offered in academic circles.

If we start by considering the horizontal diagonal as the centre of the square, we can see that it is made up of two isolated triangles, one of which points towards the heavens while the other points downwards. However, in its spiritual dimension, Mvet Ekang does not relate to earthly realities, which are more cellular, so only the trilateral figure pointing towards the heavens is taken into account. According to the symbolism of the triangle, the one pointing towards the heavens denotes man, the sun, fire, the active, etc,. In view of the above, the greatness of sacred art is still justified according to the symbolism of the triangle, outside the traditional perception described above. Everything seems to indicate that modernity is merely a gradual explanation of the discoveries of other times. There is no such thing as a new science, just new discoveries from different angles. The greatest science of all is the science of self-knowledge.

51

Using the vertical line as the centre of the geometric figure, we always obtain two isolated triangles, one pointing to the right and the other to the left. The symbolism of the triangle remains the same, because the right-hand side is often considered to be the place of the chosen ones, of those who have known or are knowing success, while the left-hand side refers to weakness, betrayal, etc. The passive and the active are always put together to express the balance of forces. The moon and the sun are opposites in the sense that one is heat, while the other is gentleness, yet humanity could not survive without them. This is a complementary opposition, not a harmful one. The universal forces complement each other because each has its own moment of manifestation; life cannot be limited solely to perfection or negativity. The unconditional exclusion of one or the other is not relative to human nature, but rather the work of the spirit. Perfection is a revelation of the spiritual work, you have to die to know a world without wars, family conflicts, etc,. The best we can do is not to reject evil as a non-integral reality of our lives, but rather to know how to overcome it, which is where the notion of balance comes in.

The pyramid itself is a reminder that evil will not be wiped out by the hand of man, but rather by a higher intelligence. In simple terms, nature always knows how to take care of itself. The return to square one is a long march that has been marked out for a long time, long before the advent of humanity on earth. The echo produced by the cordophones of sacred art is a word manifested through vibration, from which all things are born. Water is made fertile by the echo driving its undulatory movements, so that the pyramid is born.

5-) the geometric shapes of the pyramid

a-) the right angle

It has been noted that the square used to illustrate the inverted triangles above adopts a vertical posture, while resting its construction on the diagonals. These diagonals have an arithmetical configuration through the 555 triplet, which is normally a mark of stagnation. However, since we're talking about a vertical ascent, it's worth reviewing the configuration of the right angle. The point here is to highlight another feature of the right angle, taken from the perforations in the Mvet instrument. Let's look again at the right angle observed above:

A right angle

~
~
~
~
~

~ ~ ~ ~ ~ ~ ~ ~ ~ ~

Source: NNANG EBANE Sosthene Tresor

There are, of course, 5 perforations on the left-hand side, 5 perforations on the bridge, and 5 perforations on the right-hand side, however, the bridge itself is inserted within another perforation known as the centre perforation. So the right angle will now appear with an extra dot in the centre as follows.

A right angle

Source: NNANG EBANE Sosthene Tresor

By inserting an additional point at the centre of the right angle, we go from triplet 555 to triplet 666, however, the arithmetical triplets taken in this formation do not evoke ascension in a practical way or in such a way that ascension can be observed with the naked eye. In this case, the total number of horizontal points should be counted, then their centre of arithmetical symmetry should be found, above which exactly the same number of points should be added, making up either its left or right half. T(n)=n+1+n equals T (5)=5+1+5=11. The number 1 in the centre represents neutrality in the regression triplet, which will be transformed into the number 6 in the ascension triplet. T(n)=n+(n+1)+n is equivalent to T (5)=5+(5+1)+5=5+6+5=16. The number 6 symbolises both the height of the triangle and its vertex, while the numbers 5 on each side designate both its ascending and descending sides, and the number 16 refers to its base. We discover that the homogeneous triplet, 555 or 666, is certainly a perception of the triangle, but does not really include the base.

b-) right angles and rectangles

The previous point was based on the arithmetical configuration of the right angle according to the configuration of the perforations on the sacred Mvet Oyeng instrument. We now need to start from the right angle and work our way up to the rectangle. It's just a question of reconsidering the right angle, then completing the whole plan by adding 5 extra points on each horizontal perforation. Let's reconsider the second right angle:

A right angle
Source: NNANG EBANE Sosthene Tresor

Let's fill it in so that it has a completely homogenous shape, as shown below:

Source: NNANG EBANE Sosthene Tresor

The rectangular sketch obtained is 6 cm wide and 11 cm long, starting from the Mvet Ekang made up of 5 cordophones. We can see that 25 points are added on each side of the right angle to obtain the rectangle. There are a total of 66 points making up the plane, because bxh or hxb equals 11x6 or 6x11, or 25+25+16=(2x25)+16=50+16.

In addition, we can see that the number taken for the width and length of the sides of the rectangle provides square-shaped spaces whose corresponding numbers are smaller than these by a small unit. We know that the length is 11cm, so the internal length is 10cm, and that the width is 6cm, so the internal width is 5cm. We discover that the triplet 555 relates not only to the triangle, but also to the tripartite perception of the internal width of the rectangle, corresponding to the organisation of 24 points on the plane, i.e. the number 6666. The number 555 and the number 6666, although related to the number 7 as a unit and added together, define the internal and external parts of the rectangle. From an internal point of view we have: length 5 and width 3, while from an external point of view: length 6 and width 4. So, 3x5=15 gives the triplet 555, and 4x6=24 gives the number 6666.

Returning to the composition of the arithmetical triplets, we first note that: the ascension triplet 565 becomes 65656, then the regression triplet 515 becomes 65156, finally the homogenous arithmetical triplets 555 becomes 6565656 and the triplet 666 becomes 6565656 . It should be noted that the identical arithmetical triplets reflect a very particular reality, relating to the unity of the number 3 with the number 4. In fact, the former is present within the latter in the form of an interval, while the former relates to the boundaries of the latter. According to the rectangle, the number 6 symbolises the boundaries of the intervals, while the number 5 represents the interval. So, if the number 5 appears 3 times, the number 6 appears 4 times, hence their singular appearance relates to the number 7, as the unity of the triangle with the square. It should be remembered that the Sacred Harp, traditionally called Ngoma or Ngombi, offers exactly one tripartite reading of the number 8, and one quadruple reading, hence the numbers 888 and 8888. The numbers counted in a singular way also rest on the number 7, with 888 symbolising the number 3 and 8888 defining the number 4, so 3+4=7. It is clear from the above that the triangle is inscribed within the rectangle.

c) rectangles and triangles

The study of the Ngombi and Mvet Ekang instruments seems to show that the pyramid abounds in a multitude of geometric figures. In line with this idea, the triangle can be seen in the rectangular sketch below:

Fig: the inscribed triangle within the rectangle

Several colours have been used to highlight the trilateral figure in this quadrilateral. First of all, it should be noted that as the plane is made up of 66 points, only 36 are used to construct the triangle, as there are 15 black points on either side of the triangle. There are 5 cordophones, so in the same vein 5 triangles are compiled for a faithful rendering of the sacred art. It is only right that this should be clearly demonstrated:

Fig: the inscribed triangle within the rectangle

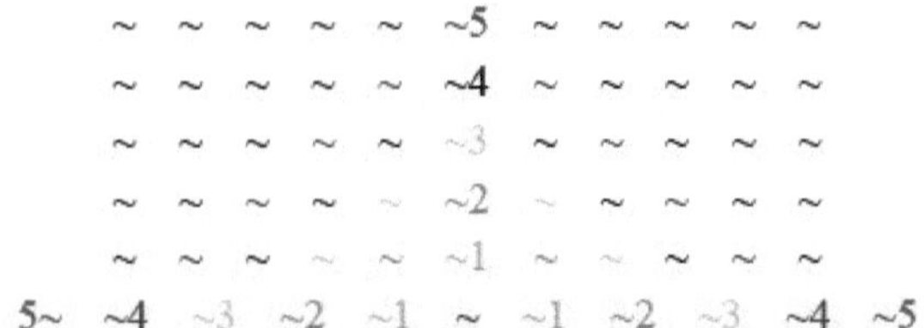

Obviously, we obtain 5 triangles in the light of what has gone before through the triplets 111, 222, 333, 444 and 555. The latter clearly show that sacred art is derived from this mathematical confirmation. It is also possible to read the triangle in a different way, by taking into account the length of the sides in relation to its base. There are 6 cm of sides for 11 cm of base, 5 for 9, 4 for 7, 3 for 5, and 2 for 3. Each triplet of regression is therefore related to the size of the base, because we have: 5+1+5=11, 4+1+4=9, 3+1+3=7, 2+1+2=5, and 1+1+1=3. The length of the sides is then related to half the regression triplet plus the digit or number taken as the centre of arithmetic symmetry. 1+5=6, 1+4=5, 1+3=4, 1+2=3 and 1+1=2, hence the triples 565, 454, 343, 232 and 121 are described as ascending. The number 5 therefore denotes the interval made practical by the action of the cordophones. The following demonstrates the clear perception of the external number of civilisational memory:

Fig: the inscribed triangle within the rectangle

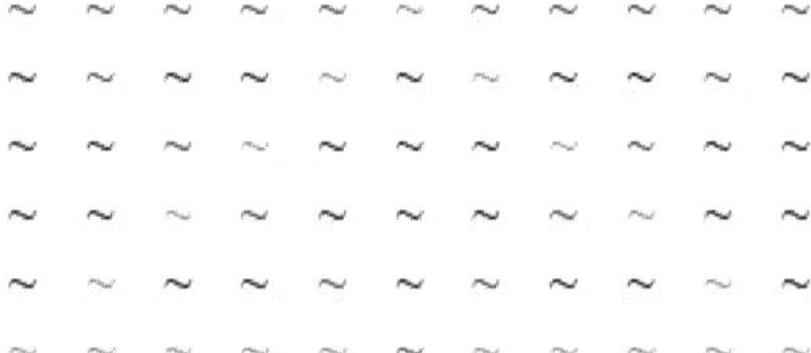

Source: NNANG EBANE Sosthene Tresor

By inserting a non-zero value in the centre of the digits, we obtain three parts of numbers ranging from 1 to 6, at which point the digit 5 reverts to its interval value. However, there are still 5 arithmetic triplets: 666, 555, 444, 333 and 222, or 222, 333, 444, 555 and 666. We can see that the number 1 is a neutral unit in the counting of arithmetic triplets.

In the light of the above, Mvet Ekang evokes the word of the invisible through the sum of the interval, which we know but which matters little. It puts forward the idea that the invisible animates the visible, or that the invisible is hidden within the visible. It is from this perspective that mathematics, through the writing of numbers and geometrical figures, is a perfect transposition of the close relationship between the human soul and its body, under the impulse of a supreme being who is the author of this mysterious work, representing himself in the combined form of echo and vibration. Mvet Ekang advocates unconditional recognition of the soul as subject to a certain constant evolution. The soul must evolve.

d-) the rectangle and crossed triangles

Until now, the rectangle has been a geometric figure used as a support for other figures. We should continue in the same vein, by including crossing triangles. Crossed triangles are defined as two triangles of the same nature whose vertices point towards each of their bases. If one points downwards, then the other points upwards. The following is a perfect illustration of this observation:

Fig: crossed rectangle and triangle

Source: NNANG EBANE Sosthene Tresor

The rectangle above does indeed show the image of two triangles, with their vertices pointing towards opposite bases. Unexpectedly, we also see the presence of a square, as a result of their inversion. The whole image reveals not just two intersecting triangles, but also a square surrounded by four triangles of the same type. This is how an open pyramid is created, by extending the plane along the right angle of sacred art. In the end, Mvet Oyeng offers several images of the pyramid. It is useful to present the entire plan, even though the sacred art relates exclusively to the upper region. It would probably not be a mistake to propose the following plan:

Fig: the pyramid and its facets

```
#  #  #  #  #  €  #  #  #  #  #
#  #  #  #  €  #  €  #  #  #  #
#  #  #  €  #  #  #  €  #  #  #
#  #  €  #  #  #  #  #  €  #  #
#  €  #  #  #  #  #  #  #  €  #
€  #  #  #  #  #  #  #  #  #  €
#  €  #  #  #  #  #  #  #  €  #
#  #  €  #  #  #  #  #  €  #  #
#  #  #  €  #  #  #  €  #  #  #
#  #  #  #  €  #  €  #  #  #  #
#  #  #  #  #  €  #  #  #  #  #
```


\#   The right angle, which was initially taken to be the very skeleton of the triangle when mathematical analysis was introduced using the traditional instrument, eventually turned out to be the axis of symmetry and the median of the square in the light of the foregoing. From the result obtained, we can deduce that two rectangles joined at the top and bottom form a square. In this situation, the square is 11 points or 11 cm long.

\#   The diagonals are the same length as the sides of the tile, i.e. 11 points against 11 points. The diagonals and medians or axes of symmetry all intersect at the centre of the tile.

\#   The 11 cm square is used as the base figure for the set of geometric shapes and figures inside it. It occupies a usual or normal posture.

€: A second edge appears within the latter in a vertical position. It measures 6 points or 6 cm in length.

\#   The diagonals of the ordinary square become the axes of symmetry or the medians of the second square.

\#   Inversely, the axes of symmetry or the medians of the ordinary square become the diagonals of the vertical square.

6-) arithmetic analysis

a-) the ordinary edge

From the above result, we can see that the civilisational memory has an ordinary square within it, whose side lengths relate to an odd number. This number is chosen not only to trace the length of the sides, but also to highlight its axes of symmetry. This is why the trilateral figure that makes up its skeleton has the angular unity of horizontal and verticality. However, it is the verticality that is represented by half, in order to remain faithful to the excedent value of the Mvet expression. The number 11 has the number 1 as its arithmetical centre of symmetry, taking the number 5 as its symmetrical images, so 5+1+5=11. The expression of the surplus value is then constructed by uniting the centre with one of the symmetrical values, giving: 5+(5+1)+5=5+6+5=16. The number 6 now symbolises the height of the triangle, but its value is not clearly shown on the object's skeleton.

b-) the vertical edge

It can also be seen that the posture of the vertical edge within another so-called normal edge is achieved by the odd nature of the length of the first edge. The second or vertical square is a quadruple view of the central excedantaite value of the number 11. This is why it measures 6 points or 6 cm in side length. The quadrilateral of length 6 cm is inscribed within the quadrilateral of length 11 cm, which is why the numerical triplet in general appears to be a simplified form of writing of this geometric theorem. It is thus possible to start with the triangle, passing through the rectangle, and finally arrive at the square, as the basic geometric figure.

c-) the nature of the digital triplet

The arithmetical triplet itself evokes the idea of compiling several geometric figures, starting with the triangle as the primary figure. In the light of the above, the number 3 is revealed as the very expression of multiplicity within unity. But it is a multitude that wants to be united both horizontally and vertically, in other words, by filling the universe in its totality. Among the triplets, there are the following chifes and numbers: 3,5,7,9,11,13,15,17,19, etc,.. They allow us to obtain the vertical posture of the square as one of its main characteristics, centred around the transformation of the axes of symmetry into diagonals and of the diagonals into axes of symmetry, in relation to the inscription of the vertical square within the ordinary square.

Even numbers can also be used as arithmetic triplets, but as they have a centre of symmetry other than 1, the triangles, axes of symmetry and diagonals will appear twice.

7-) the pyramidal alphabet

a-) Mvett and fractals

It is in the natural order of things always to remain at the heart of the organisation of sacred art, in order to bring out another particularity relating to the art of representation. When we studied the dual external and internal structures of the Mvet instrument, we realised that they could vary continuously from a lower to a higher stage. In fact, the strings of the Mvet are arranged in ascending order from bottom to top, while in descending order from top to bottom, they are similar to triangles that follow one another, so that the whole can be likened to the steps of a ladder or staircase. The quality of the Mvet Ekang instrument once again plunges our analysis into the heart of mathematics. This construction process is known as fractal geometry:

"A fractal figure is a malignant object with a structure similar to all scales. It is an "infinitely fragmented" geometric object whose details can be observed on an arbitrarily chosen scale. By zooming in on a part of the figure, it is possible to find the whole figure; it is then said to be "self-similar".

This passage proves once again that the Mvet is not only a religious instrument, as many people think, but also that it is above all a mathematical tool, which must be studied in depth in order to learn more from the treasures it abounds in. So it's not surprising that it should be proposed as an object of mathematical study in the academic world. It was shown above that the notion of fractal geometry is built around an algebraic organisation, i.e. the arithmetic triplet. The link between the trestle and the horizontal branch, which in an advanced analysis are the medians of an ordinary square, and the diagonals a vertical square inscribed within the latter. It should be noted that this notion of mathematics is also present within nature itself, once again attesting to the qualification of Mvet Ekang as the expression of multiplicity within unity, or defining it as a palpable reduction of universal facts into a practical object possessed by man. We can illustrate this as follows:

Figure: Romanesco cabbage

https://www. Wikipedia.org<wiki>fractale/2021(le8juillet)

If you look closely at this photograph, you can see that the same image is reflected spectacularly across the entire surface of the cabbage. The smaller plant formations are above the larger ones. However, on the Mvet instrument, the shorter strings are located below those of greater length, the opposite of the plant distribution seen in this photograph. This perception is obviously possible when the side branch points downwards. Like the romanesco cabbage, the fern is also arranged according to fractal geometry. The image below is a perfect illustration:

Figure: A fractal fern

These different photographs clearly show that the geometric shape is reflected in every dimension of the plane on the object under study. As far as the fern is concerned, we can count about ten instant reproductions on each side, so a whole sheet will have twenty representations in total, as an approximate value. However, it is difficult to assign an approximate value to romanesco cabbage. With the same idea in mind, we studied the internal structure of the three-string Mvet as a basic object of study, given the predominant repetition of the number 03 on the instrument, in order to discover the number of symbols it contains. Fortunately, we were able to obtain countless unitary, binary and trinary symbols.

b-) Pyramidal alphabet board

The board offers a gradual and consecutive growth of triangular symbols in three categories: unitary, binary and trinitary. It is possible to go much further, but we know that we will never be able to bring out all the facets of the object, so we will be content with a minimalist but interesting analysis. The following symbols are absolutely magnificent:

Sheet: Pyramidal Alphabet

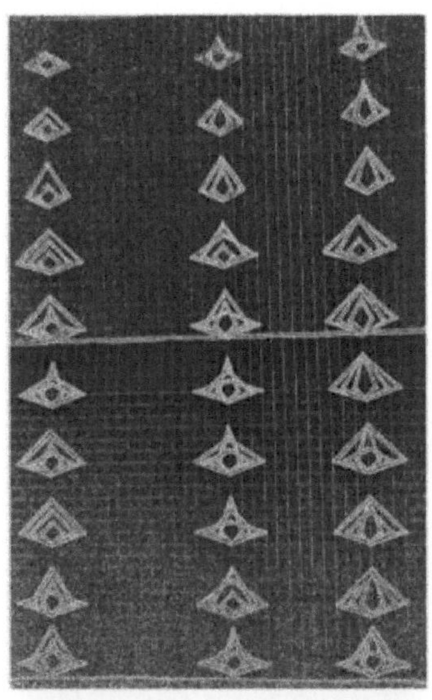

In view of this, it is clear to see that the symbols vary from the smallest unit to a larger one. The growth of the symbols reflects the idea that it would be judicious to start from nothing in order to achieve exceptional results. It is important to emphasise, however, that the growth of symbols reveals the presence of the human eye, before merging with the flame and the stars. Everything seems to link man to the sky or to the sight of the luminous stars, and we have to believe that this alphabet defines the spiritual greatness of the human being who is anxious to master the workings of the luminous stars, either to situate himself in relation to a geographical region or to solve a riddle. This is how the human eye ends up metamorphosing in order to merge with the solar planet, with a view to having a panoramic or royal view of the universe. The pyramidal alphabet defines man as a being in search of knowledge, from which it is by no means excluded that he will merge with nature. The human eye must therefore merge with the luminous stars of the sun and moon, or with the stars, the expression par excellence of spiritual awakening. However, it is necessary to determine the number of symbols, which can be obtained within the 3-string Mvet, if the string is counted singularly from left to right, or the 6-string Mvet, if each portion on the left or right is counted independently.

61

Before proceeding with a statistical study of the symbols of the pyramidal alphabet, it is only natural to be able to suggest the uses to which they can obviously be put. It would not be wrong to keep repeating ourselves, to get carried away by the mysterious contours of the sacred arts, of which the Mvet Ekang is currently the subject. So as not to stray from the subject of mathematics, we will just propose Mfine Nzalang, the Thunder Shield, congregated in honour of Lord Ekang Oyono Ada Ngoine.

Fig: symbols and the art of representation

Figure: Mfine Nzalang Oyone Ada Ngoine

✳✳✳

(Unitary, binary and trinitary symbols<183 +1>)

Sacred art plunges the observer into a fantastic universe where there is no need to think to understand it, just to admire its marvellous decor and write it down faithfully on a support. Knowledge is not man's property, he just uses it in a non-permanent way. It would therefore be foolish to talk about Mvet without mentioning Lord Oyono Ada Ngoine, as this is simply the result of a testimony. There are many other symbols, but this Ekang shield will do.

8-) the numerical values of symbols

The symbols making up this pyramidal alphabet obey specific numerical values. In other words, each symbol has its own specific coordinates. To this end, it reveals a triune organisation of symbols, the triplet 123. So we start with a unitary symbol, then move on to a higher, binary symbol, and finally to a third, trinitary symbol. It has to be said that trigular forms grow from 01 to 02, then from 02 to 03, and from 03 to 04, and even more. Growth and decay are two unified orders that always accompany these different symbols.

a-) Unit symbols

Unit symbols do not have a fractal geometry, they do not contain a lining, like the first three symbols: 1; 2; and 3, which have a symbol that evolves in three dimensions. The symbol below is a perfect illustration of this argument:

Figure: Akong

(The lance)

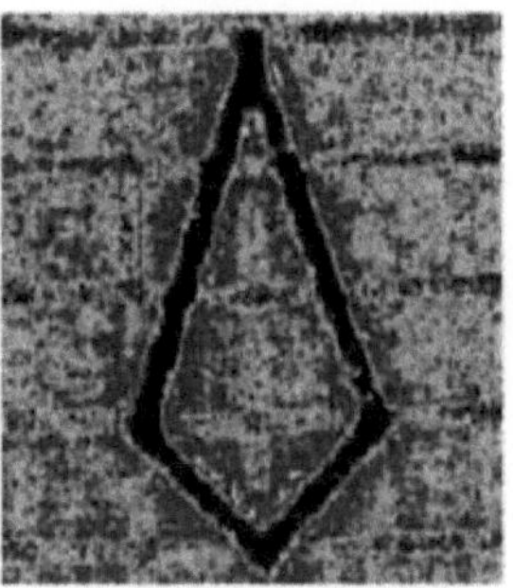

Source: NNANG EBANE Sosthene Tresor

Akong is a unitary symbol, with co-ordinate: (1-3). It has a square base and a triangular height. It is the highest degree of the unit symbols, those it precedes having coordinates (1-1) and (1-2). The central calabash is taken as the point of origin of each symbol, because it is necessary to maintain the logic according to which it is the triangle that gives height to its square base.

b-) Binary symbols

They contain duplicates, but in the order of two, like second symbols. There are four types of bianary symbol: combined, double combined, static and static combined.

b-1) combinations:

These are symbols obtained from coordinates that vary on a scale of 1 to 3. Symbols vary according to the numerical magnitude assigned to them, but in order of two, squares and triangles together.

Numerical table of binary symbols

1			2		
1-1	1-2	1-3	2-1	2-2	2-3
2-1	2-1	2-1	1-1	1-1	1-1
1-1	1-2	1-3	2-1	2-2	2-3
2-2	2-2	2-2	1-2	1-2	1-2
1-1	1-2	1-3	2-1	2-2	2-3
2-3	2-3	2-3	1-3	1-3	1-3
3	3	3	3	3	3

Source: NNANG EBANE Sosthene Tresor

This statistical table shows the number of possible combinations to obtain binary symbols. There are a total of 18 binary symbols, i.e. 9 combinations for each digit 1 and 2. So 9+9=18 or 9x2=18. The binary symbol below effectively links two triangles at the higher level:

Figure: Ayem

(Knowledge)

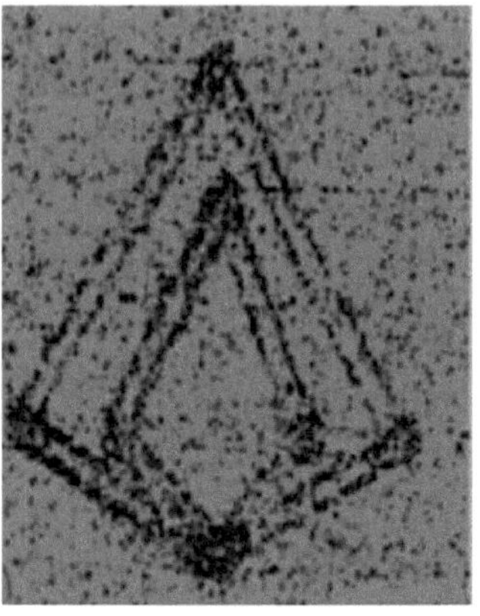

Source: NNANG EBANE Sosthene Tresor

This is a binary symbol. It comprises the following strings: (1-2 ; 2-3). There is no doubt that the inner triangle is smaller than the outer one, thus expressing the notion of fractal geometry, of a figure inscribed within another of the same nature.

b-2) combined doubles:

They are so called because they are obtained from the cross product of the previous values.

Numerical table of binary symbols

1-2	1-2	1-2	1-1	1-2	1-1
1-1	2-1	3-1	1-2	2-1	3-2
2-2	2-2	2-2	2-1	2-1	2-1
1-1	1-1	3-1	1-2	2-2	3-2
3-2	3-2	3-2	3-1	3-1	3-1
1-1	2-1	3-1	1-2	2-2	3-2
3	3	3	3	3	3

Source: NNANG EBANE Sosthene Tresor

Like the previous statistical table, it also presents 18 possibilities. That's 9 symbols for each number in the plan, 1 and 2. The table is linked from top to bottom.

b-3) statics:

Each time one triangle-square takes on a value of one unit, the second increases while retaining the same value linked to the order of revolution. This is an overall evolution and not an independent one, as has been observed previously. The image of the whole can be that of a pregnant woman, because the child is the new element, which is naturally internal to her.

Numerical table of binary symbols

1			2		
1-1	1-2	1-3	2-1	2-2	2-3
2-1	2-1	2-1	1-1	1-1	1-1
1-2	1-2	1-2	2-2	2-2	2-2
2-2	2-2	2-2	1-2	1-2	1-2
1-3	1-3	1-3	2-3	2-3	2-3
2-3	2-3	2-3	1-3	1-3	1-3
3	3	3	3	3	3

Source: NNANG EBANE Sosthene Tresor

The statistical table shows that there are 18 possibilities. That's 9 possibilities for each of categories 1 and 2. The symbol below illustrates this observation perfectly:

Figure: Dzi Ayem

(The Inquisitive Eye)

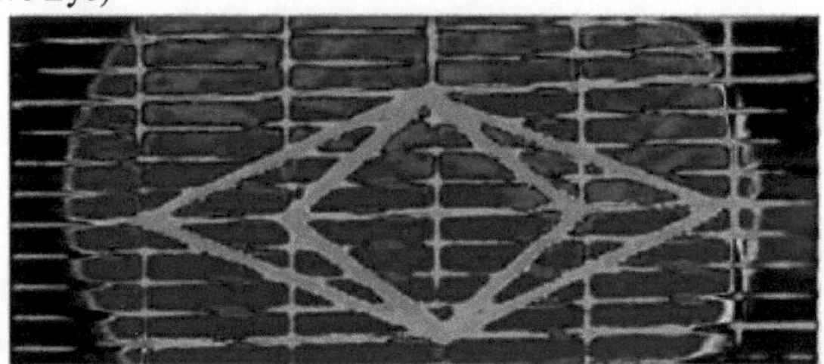

Source: NNANG EBANE Sosthene Tresor

If we look closely at this symbol, we can see that it is similar to the human eye. Both the outer and inner triangles have the same upper vertex, which makes this symbol one of the combined static symbols. It expresses sharing, unity, working together and brotherhood,
love. Isn't it said that there's strength in numbers?

b-4) combined statics

They are obtained by investing the different strings of the previous symbols. All you have to do is exchange the higher values with the lower ones, so that the inner triangle is greater than the outer triangle. The following statistical table is a striking example:

Numerical table of binary symbols

1-2	1-2	1-2	1-1	1-1	1-1
1-1	2-1	3-1	1-2	1-2	3-2
2-2	2-2	2-2	2-1	2-1	2-1
2-1	2-1	3-1	2-2	2-2	2-2
3-2	3-2	3-2	3-1	3-1	3-1
3-1	3-1	3-1	3-2	3-2	3-2

3	3	3	3	3	3

Source: NNANG EBANE Sosthene Tresor

These different tables show that each pair of four digits constitutes the coordinates of a symbol. The table should be read vertically from top to bottom. There are a total of 18 symbols in this set. The symbol below is a perfect illustration:

Figure: Dzi Ayem

(The Inquisitive Eye)

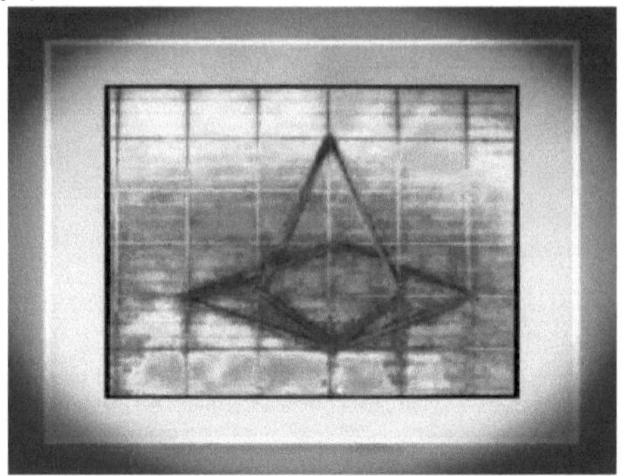

Source: NNANG EBANE Sosthene Tresor

The symbol above has co-ordinates (1-3; 2-1), so it shows that the inner triangle rises above the square, while the square (outside) remains stable. This correspondence effectively describes a triangle rising above the square, in the image of a side face of the pyramid. As far as the human eye is concerned, we can see that the square is symbolic of the orbit, while the triangle indicates the pupil emerging from the orbit. This symbol seems to indicate that mankind needs to detach itself from earthly reality in order to contemplate the celestial world, which guarantees absolute truth. It is the expression par excellence of spiritual greatness.

c-) Trinitarian symbols

They are made up of three triangles that can constantly change order, either the first keeping a value of 1(1-1), the second a value of 2(2-2) and the third a value of 3(3-3), or these values can be reversed at any time, so that the first has a value of 3(1-3), the second a value of 1(2-1), the third a value of 2(3-2), and so on. It has to be said that this numerical rotation shows that there is no evolution within the universe. In fact, the same elements follow one another, just like the days of the week and genealogy. It is this monotony that drives Ekang thinkers to seek a single world, with no variation whatsoever. Hence Mvet is neither past, present nor future, but is. The following statistical table gives an overview of the numerical values of the Trinitarian symbols:

c-1) tripartite symbols

The three triangles have nothing in common, apart from their lower centre. However, each one has a particular pitch value, in line with the stringophones of the Mvet Oyeng instrument.

66

Numerical table of Trinitarian symbols

1		2				3		
1-1 2-2 3-3	1-2 2-1 3-3	1-3 2-1 3-2	2-1 1-2 3-3	2-2 1-1 3-3	2-3 1-2 3-1	3-1 1-2 2-3	3-2 1-3 2-1	3-3 1-1 2-2
1-1 2-3 3-2	1-2 2-3 3-1	1-3 2-2 3-1	2-1 1-3 3-2	2-2 1-3 3-1	2-3 1-1 3-2	3-1 1-3 2-2	3-2 1-1 2-3	3-3 1-2 2-1
1-1 2-3 3-3	1-2 2-3 3-3	1-3 2-3 3-3	2-1 1-3 3-3	2-2 1-3 3-3	2-3 1-3 3-3	3-1 1-3 2-3	3-2 1-3 2-3	3-3 1-3 2-3
3	3	3	3	3	3	3	3	3

Source: NNANG EBANE Sosthene Tresor

This table can be read vertically, from top to bottom. The main aim of this study is to provide a comprehensive overview of the various symbols that can be obtained from the internal structure of the Mvet. It is also important to study the numerology of the Mvet in order to understand it and compare it with other religious instruments. We have a total of twenty-seven symbols: 3*9=27 or 9+9+9=27. The symbols shown include triangles in a fractal posture. The symbol below is a perfect illustration of this assertion:

Figure: Bibulu

(A woman's sex)

Source: NNANG EBANE Sosthene Tresor

Its coordinates are the following numerical values: (1-1;2-2;3-3). It is a Trinitarian symbol. It is a faithful reproduction of the arrangement of the strings on the Mvet instrument. Note that the first triangle is less prominent than the second, and that the second is also less prominent than the third. The set of triangles depicts a simultaneous evolution, a zone where waves are emitted on a radar screen, or the curves indicating the volume level on an amplifier.

c-2) combined doubles:

These symbols are almost identical to those above, except that the coordinates can be reversed, but they always relate the triplet 123. The following table is a clear example of the above:

Numerical table of Trinitarian symbols

1			2			3		
1-1 3-3 2-2	1-2 3-3 1-2	1-3 2-3 1-2	2-1 3-3 2-1	2-2 3-3 1-1	2-3 1-3 2-1	3-1 1-2 2-3	3-2 1-2 3-1	3-3 2-2 1-1
1-1 2-3 3-2	1-2 1-3 3-2	1-3 1-3 2-2	2-1 2-3 3-1	2-2 1-3 3-1	2-3 1-3 2-1	3-1 3-2 2-1	3-2 3-2 2-1	3-3 1-2 2-1
1-1 3-3 3-2	1-2 3-3 3-2	1-3 3-3 3-2	2-1 3-3 3-1	2-2 3-3 3-1	2-3 3-3 3-1	3-1 3-2 3-1	3-2 3-2 3-1	3-3 3-2 3-1
3	3	3	3	3	3	3	3	3

Source: NNANG EBANE Sosthene Tresor

This table can be read vertically, from top to bottom. The main aim of this study is to provide a comprehensive overview of the various symbols that can be obtained from the internal structure of the Mvet. It is also important to study the numerology of the Mvet in order to understand it and compare it with other religious instruments. There are a total of twenty-seven symbols in each table, i.e. 3*9=27 or 9+9+9=27. The symbol below is a good illustration of the above:

Figure: Awoum (The river)

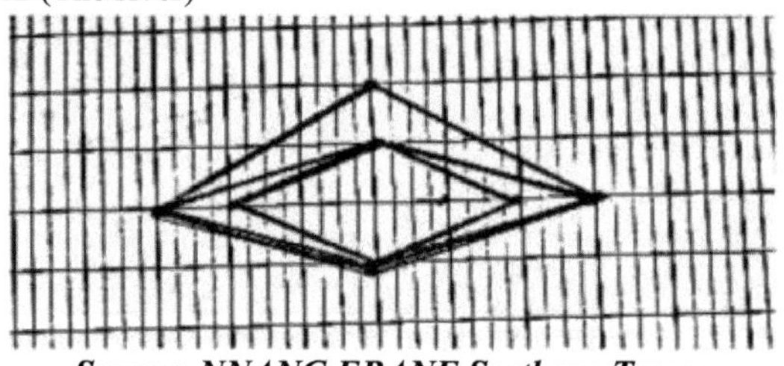

Source: NNANG EBANE Sosthene Tresor

This symbol combines an eye with a triangle rising above it. The whole also reveals a pyramid with a square base, the height of which is erased to allow a luminous view of its base. There is no doubt that the symbols of the pyramid alphabet are designed to bring the human eye to the physical structure that is the pyramid. It therefore symbolises a supreme view of the universe, the ideological conquest of power. This symbol has the coordinates (3-3; 3-2; 1-1).

c-3) statics:

They are obtained by adding all the triangles together at a main vertex. In other words, the inner and outer triangles share a common vertex. The 123 triplet is expressed unanimously through these different symbols. The following statistical table illustrates this point:

Numerical table of Trinitarian symbols

1	2					3		
1-1 2-1 3-1	1-2 2-1 3-1	1-3 2-1 3-1	2-1 1-1 3-1	2-2 1-1 3-1	2-3 1-1 3-1	3-1 1-1 2-1	3-2 1-2 2-1	3-3 1-1 2-1
1-1 2-2 3-2	1-2 2-2 3-2	1-3 2-2 3-2	2-1 1-2 3-2	2-2 1-2 3-2	2-3 1-2 3-2	3-1 1-2 2-2	3-2 1-2 2-2	3-3 1-2 2-2
1-1 2-3 3-3	1-2 2-3 3-3	1-3 2-3 3-3	2-1 1-3 3-3	2-2 1-3 3-3	2-3 1-3 3-3	3-1 1-3 2-3	3-2 1-2 2-3	3-3 1-3 2-3
3	3	3	3	3	3	3	3	3

Source: NNANG EBANE Sosthene Tresor.

This table can be read vertically, from top to bottom. The main aim of this study is to provide a comprehensive overview of the various symbols that can be obtained from the internal structure of the Mvet. It is also important to study the nurmerology of the Mvet in order to understand it and compare it with other religious instruments. There are a total of twenty-seven symbols in each table: 3x9=27 or 9+9+9=27. The following symbol is a perfect illustration:

Figure: Elam ye ngnol
(The light of the body)

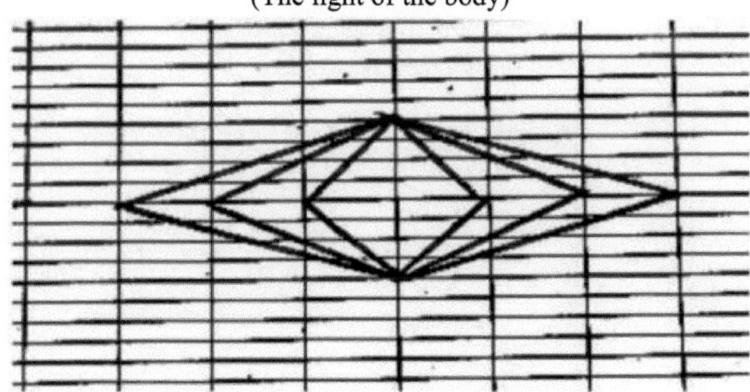

Source: NNANG EBANE Sosthene Tresor.

This symbol, called Elam Ye Ngnol, is undoubtedly a Trinitarian symbol, but the three triangles share a common vertex. Its coordinates are 1-1, 2-1, 3-1. It corresponds to an open eye, so it symbolises the light of the body, as its name suggests. In reality, it is a regression of the three triangles into squares. The first is inscribed within the second, and the second is inscribed within the third. They are arranged in ascending and descending order.

It was shown above that the combined static symbols involve the union of three triangles, only two of which share a common vertex. However, the present situation is specific in that it focuses on a triangle capable of dividing into three dimensions. The following statistical table illustrates this particular point:

Numerical table of Trinitarian symbols

1			2					
1-1 1-3 1-2	1-2 1-3 1-2	1-3 1-3 1-2	2-1 1-3 1-1	2-2 1-3 1-1	2-3 1-3 1-1	3-1 1-2 1-1	3-2 1-2 1-1	3-3 1-2 1-1
1-1 2-3 2-2	1-2 2-3 2-2	1-3 2-3 2-2	2-1 2-3 2-1	2-2 2-3 2-1	2-3 2-3 2-1	3-1 2-2 2-1	3-2 2-2 2-2	3-3 2-2 1-2
1-1 3-3 3-2	1-2 3-3 3-2	1-3 3-3 3-2	2-1 3-3 3-1	2-2 3-3 3-1	2-3 3-3 3-1	3-1 3-2 3-1	3-2 3-2 3-1	3-3 3-2 3-1
3	3	3	3	3	3	3	3	3

Source: NNANG EBANE Sosthene Tresor

This table can be read vertically, from top to bottom. The main aim of this study is to provide a comprehensive overview of the various symbols that can be obtained from the internal structure of the Mvet. It is also important to study the numerology of the Mvet in order to understand it and compare it with other religious instruments. There are a total of twenty-seven symbols in each table: 3x9=27 or 9+9+9=27. The symbol shown below illustrates this assertion perfectly:

Figure: Aba Medzo

(The guardhouse)

Source: NNANG EBANE Sosthene Tresor

The symbol above shows a fragmented pyramid with a square base. In fact, combined trinity symbols can have a single angular origin or different angular origins, but two triangles can share a common vertex. In this case, the angular origin is common to a single triangle with different vertices or trine vertices. All these symbols show that the pyramid has several forms, and that the Mvet instrument is an art that transcends itself. This shows once again that the Mvet instrument is perceived as a fixed reality when viewed from the outside, but the mobile nature of the strings expresses itself far beyond its simple external perception. It is the expression of universal realities that are both fixed and mobile.

NB: The symbols of the pyramidal alphabet are so innumerable that they would constitute an entire book in themselves. For example, we should no longer consider a common centre for the triangles, allowing each of them to be represented independently, but within a single plane. The following symbol is similar to that of the famous "Umbro" brand:

Figure: Abora
(Thanks)

Source: NNANG EBANE Sosthene Tresor

The Abora symbol shown above is a binary symbol, representing a small square inscribed within another. This symbol is strangely similar to both the human eye and its mouth. As far as the pyramid is concerned, it naturally refers to its base, with no connection to the triangles. It is clear from this that the Mvet contains countless marvels that even this book cannot contain, but we can be satisfied with what is offered here.

9-) the static overview of symbols

This is a more detailed presentation of the study based on the internal organisation of the Mvet Ekang instrument. It should be remembered that this is an instrument made up of 3 cordophones, which is thus taken as an illustration, before pretending to project onto those which contain more. It is therefore a statistical summary that is offered here.

Summary table of contact details

	Unit symbols	Binary symbols	Trinitarian symbols

	1	18	27
	1	18	27
	1	18	27
		18	27
Number of appearances	3	4	4
		9-9	9-9-9
		9-9	9-9-9
		9-9	9-9-9
		9-9	9-9-9
Number of appearances		8	12
		3+3+3+3	3+3+3+3+3+3+3+3+3
		3+3+3+3	3+3+3+3+3+3+3+3+3
		3+3+3+3	3+3+3+3+3+3+3+3+3
		3+3+3+3	3+3+3+3+3+3+3+3+3
Number of appearances		16	36

Source: NNANG EBANE Sosthene Tresor

According to this statistical table, there are three (3) "unit" symbols, i.e. 3x1=3, seventy-two (72) "binary" symbols, i.e. 18x4=72, and one hundred and eight (108) "trinity" symbols, i.e. 27x4=108. In total: 3+72+108=183 symbols. It should also be noted that the numbers appearing as symbols evolve from 3 to 4, just as the base of the pyramid evolves from a triangle to a square. In this way, we can see that the number 4 is the very expression of constant and stagnant revolution. This is why we speak of the "numerical stagnation of growth". Hence: 4+4=8+4=12 +4= 16+4=20+4=24+4= 28+4=32+4=36. In view of the above, the number 4 is the guiding unit of this numerical condensation. It symbolises, of course, the four cardinal points that allow us to situate ourselves geographically. But over and above this, we can obviously see that man is always keen to find his place throughout his life: finding the right job, the right woman, the right friends, quality TV series, the drink of choice, and so on. It has to be said that the number 4 shows that the human race is constantly travelling through its many variations. Our earthly and spiritual lives place us on a path leading to places that we cannot name for lack of knowledge.

This table is also useful for calculating the number of symbols that a Mvet instrument can hold, based on the number of perforations it has. Note that the instrument has both lateral and horizontal perforations, so that the horizontal value is double the lateral value. Hence the name h for the horizontal value and v for the lateral value. We set $h = (n*2)+1$, while $v = h - n$, where n refers to the number of perforations in the Mvet. If an instrument has 18 perforations, and we want to find the number of symbols it contains, we set: $h =(18*2)+1=36+1=37$, or 37=36+1. Hence:

	Unit symbols	Binary symbols	Trinitarian symbols	Total
level 1	1	18	18	37
grade 2	1	18	18	37
grade 3	1	0	18	19

This table shows that the trinity is obtained by adding the unity to the duality. We can see that the trinity is zero in the dual, because it has a drop unit. Hence we say:

degree1: 36+36+1=73

degree2: 36+36+1=73

degree3: 36+0+1=37

Hence: 37+73+73=183

The Mvet, which has 18 perforations, contains 183 symbols. These make up the pyramidal alphabet.

Example 2:

For the Mvet with 15 perforations, h =(15x2)+1=30+1=31, V =31-15=16. This gives: (15+1)+(15+15+1)+(15+15+1)=16+31+31=78. There are 78 symbols for the Mvet, which has fifteen (15) perforations. The number of symbols we found was 183, because we had focused on the trinity or triangle. We are still working on the same idea, which is to show that the square and the triangle are closely related, and that their numerical values are the numbers 3 and 4. To do this, we are going to draw up the number of symbols by unitary, binary, trinitary and quarto correspondence:

a-) unit symbols:

-Unit: 1

-Binary: 18

-Trinitarian: 27.

Note that the unit symbols do not correspond to the data number 1, but to each number in the three sets. Hence we have: 1+18+27=46. It is this number that constitutes the degree of revolution in each set. So 46+0=46

b-) binary symbols:

-Units: 1+1=2

-Binary: 18+18=36

-Trinitarian: 27+27=54

From this we can see that duality implies the addition of unity by itself. We thus obtain the following Trinitarian couple: 2+36+54=92. Or 46+46=92 or 46x2=92

c-) Trinitarian symbols:

-Units: 1+1+1=3

-Binary: 18+18+18=54

-Trinitarian: 27+27+27=81

From this we can see that the Trinity implies the addition of unity by itself, in three dimensions. This gives us the following Trinitarian couple: 3+54+81=138. Or 46+46+46=138 0u 46x3=138

d-) quarto symbols:

This situation defines the study of the number of symbols, from trinity symbols to quarto symbols. The examples below illustrate this function overflow perfectly.
Example1:
-Units:1+1+1= 3
-Binaire:18+18+18+18=72
-Trinitarian: 27+27+27+27=108
From this, we can see that the quarto implies the addition of unity by itself, in four dimensions. This gives us the following Trinitarian couple: 3+72+108=183.

or

Quarto symbols:

Example 1:
-Units: 1+1+1+1=4
-Binary: 18+18+18+18=72
-Trinitarian: 27+27+27+27=108
From this we can see that the quarto implies the addition of unity by itself, in four dimensions. We thus obtain the following Trinitarian couple: 4+72+108=184.
From where we pose:
Example 2:
Q=(nx3)+n-1 AN: Q=(46x3)+46-1=138+45=183.
Example2:
Q=(nx3)+n AN: Q=(46x3)+46=138+46=184, or
Q=nx4 AN: Q=46x4=184.
The examples above show that the numbers 3 and 4 merge because one implies the presence of the other. In other words, the triangle is a square in which the missing angle is not represented in a concrete way, but this does not mean that it is absent. In the same way that the square is a triangle that manifests itself through the diagonal that joins the two opposite angles. In other words, the diagonal serves as an orthogonal symmetry in a square and in a triangle with respect to the vertex of the angle to which it is opposed. In other words, the number four can be evoked and not expressed, just as it can be evoked and expressed, depending on which number is emphasised. In the first example, we are of course talking about the number 4, but the emphasis is placed on the number 3, whereas in the second case, the emphasis is placed on the number itself. Hence the total number of symbols varies between 183 and 184, depending on whether we are dealing with a triangle or a square. It should also be noted that when we count the number 4 with our fingers, there are exactly 3 intervals between them, so the number 4 is the "physical" expression of the number 3, which is "invisible". It is the latter that the instrument highlights through the number of cordophones, which is in reality a numerical interval value, in order to give expression to the invisible. Mathematics

and spirituality cannot therefore be dissociated. This calculation makes it possible to decompose numbers into orders of three and four digits, in order to facilitate their mental addition. The metamorphosis of the square into a triangle, or of the triangle into a square, is also expressed numerically, as 3+1=4 or 4-1=3.

III-) THE PUNU MASK

Sources: https : //www.art-masque-africain.com/images/2017/12/4819-orig.jpg

**A memory
reconnaissance at the location of
our late classmate
AYITO OKOU Laurence Valeska
(Lycee Public Moise NKOGHE MVE)**

1-) Observation of the vertical edge

In addition to plucked musical instruments, it has been noted that even masks incorporate mathematical features relating to the hexagram in their breasts. This representation is a compilation of numerous geometric forms generalised within Gabonese arts, and appearing just halfway through. The reflection is part of the perspective of the total reconstruction of the geometrical figure starting from its half. It is therefore essential to carry out a detailed analysis of the object of study:

"Mukuyi masks are said to represent ancestors, sometimes female. The enigmatic face of the mask is slightly triangular. Under the closed, almond-shaped eyes, swollen as if from sleep, the high cheekbones are rounded (....). The most common pattern, in the shape of scales, consists of nine lozenges".

This descriptive paragraph on religious art plunges without a shadow of a doubt into the heart of the science of mathematics, through the triangle and the rhombus. It is important to go further in this direction, while at the same time putting forward additional explanations for understanding sculptural art as a purely mathematical object. Total immersion in its characteristic features is of paramount importance.

If you look very carefully at the Punu mask, you can see that it has a square made up of 9 small internal diamonds. However, it would be interesting to understand the different contours of this geometric figure. In fact, it comprises 9 rhombuses in the shape of scales, which refer to the internal value (VI). The external value (VE) is 4, the length of the side of the tile. The method of calculation used to produce this figure consists of subtracting one unit from the external value (VE) and then calculating the result as the square (VI). The following equation is used: $n-1=x^A 2$ numerical application: $4-1=3\ 2=9.^{\pi}$

Let's take a closer look at this mathematical explanation:

- the number 5

Assume: $5-1=x\text{Л}2$ numerical application: $5-1=4\ \text{л}2=16$

- the number 6

Assume: $n-1=x\text{Л}2$ numerical application: $6-1=5\text{Л}2=25$

- the number 7

Let's say: $n-1=x^A 2$ numerical application: $7-1=6\ 2=36^{\pi}$

- the number 8

Assume: $\pi-1=x\text{Л}2$ numerical application: $8-1=7\text{Л}2=42$

- the number 9

The following equation is used: $\pi-1=x\text{Л}2$ numerical application: $9-1=8\text{Л}2=64$

In short, the external value VE of this square is 4, i.e. 4 cm long, and the internal value is 3, so the square is 9. The 9 diamond-shaped scales refer to the internal value, which is always higher than the square.

It would be difficult so far to understand why the square is represented in the centre of the mask, i.e. in the centre of the forehead, because other facts need to be added in order to do so. The approach requires an explanation in terms of the logical relationship between shapes on the one hand and numbers on the other, or between geometric shapes and figures and algebra.

2-) The triplet of ascent and construction

The vertical edge can only be constructed on the basis of the numerical triplet of ascent, the mathematical formula of which is: T(n)=n+(n+1)+n. According to any numerical application we will have: T(5)=5+(5+1)+5=5+6+5, i.e. n values of 5. Note that the centre digit is always larger than the surrounding digits by a small unit. The vertical square therefore imposes a role reversal or simply an alternation of numerical values. The so-called central value becomes outside the vertical square, while the small value (the symmetrical boundary) occupies the centre, and is always higher than the square. The extremities now occupy the centre, and the centre becomes part of the extremities: T(5)=5+6+5 equals T (5)=6+(5*5)+6 equals T(5)=6+25+6. The mathematical formula corresponding to this triplet with respect to the square is thus: T(n)=(n+1)+(n*n)+(n+1), where T(5)=(5+1)+(5*5)+(5+1)=6+25+6. It is therefore necessary to add a unit to each of the symmetrical terminals in order to correspond to the central unit, and to multiply the symmetrical terminal by its counterpart in order to obtain a new central value. In this way, we go from 5+6+5 to 6+25+6, from the triplet of the ascent to the square triplet. However, we only consider one of the symmetrical terminals and the central value of the square to define the configuration of the latter, either 6+25 or 25+6.

More explicitly, the number 6 corresponds to the length of the sides of the tile, while the number 5 defines the number of internal tiles on the tile. So if the side of a tile is 6cm long, it has 25 squares.

3-) Triplets and the rectangle

It should be remembered in passing that the numerical triplet of regression (T(n)=n+1+n) always precedes that of ascent (T (n)=n+(n+1)+n), because it allows the construction of the rectangle within which the vertical square takes shape. The second is therefore internal to the first, or the first is external to it (envelopes) the second. In practical terms, the 5+1+5 triplet is external and antecedent to the 5+6+5 triplet, which is internal. It is the so-called ascent triplet that defines the arithmetic congregation of the vertical edge, i.e.: T(n)=(n+1)+(nxn)+(n+1), hence T(5)=(5+1)+(5x5)+(5+1)=6+25+6.

The rectangle through these triplets is 11cm long and 6cm wide, while the square is 6cm long. The width of the rectangle is equal to the length of the square, hence the presence of the square in the rectangle.

The square of the Punu mask is inscribed in a rectangle measuring 7cm long and 4cm wide, so the same measurement makes the length of the sides of the square.

78

It's important to emphasise that the numeric triplets used in this layout only come from odd numbers. In the example above, it's the number 11 that comes first. All that needs to be done is to consider one of the adjacent boundaries and add it to the centre number to define it, i.e. 5+6=11.

The vertical square in which the diamond-shaped blocks of scales are found on the Punu mask is actually inscribed in a rectangle. This quadrilateral is therefore not present on the mask, probably for reasons we do not know. However, it can be argued that knowledge is not available to everyone, so you have to be curious to find out. It is a single element that is represented, and we need to start from it to discover the rest, like the African who traces his family tree from himself, i.e. from the name he bears, to his most distant ancestor. This stage of the analysis can be placed in the perspective of a hypothetical, while waiting to discover what will happen next.

4-) The arithmetical configuration of the rectangle

However, it has been observed that the sculpture no longer appears with a rhombus (vertical square) on the forehead and sides, but rather with a rhombus-shaped rectangle on the forehead accompanied by two vertical rectangles, one on each side. The central one is larger than the vertical ones, which are similar. This particular configuration is reminiscent of the central half calabash on the Mvet Ekang instrument, which is larger than the surrounding calabashes. We can see that the number of rhombus blocks is no longer the same as the square within it, but rather a number that is not based on perfect squares, i.e. 12. The association of the vertical square and the rectangle is thus made manifest as belonging to a single geometric form, without however establishing any logical relationship to it. It is therefore interesting to understand this sudden change. Let's take a closer look at the sublime photograph below:

The Punu mask

Sources: https : //www.art-masque-africain.com/images/2017/12/4819-orig.jpg

The blocks of rhombuses evolve in two orders: in order of 4 and in order of 3. We can see that the number 3 appears 4 times and the number is also represented 3 times, i.e. 3x4 or 4x3=12. If we consider them together as internal values VI1 and VI2 according to their respective dimensions, then their respective external values are VE1= 5 and VE2= 4. It is therefore a rectangle of length 5cm and width 4cm.

It's important to understand that the arrangement of the rectangle on the forehead is unique to the square. What's more, the shell-shaped diamond blocks are not in their natural position. In fact, by plagiarising 5 points against 5 on the notebook (horizontal plane) and 4 points against 4 vertically, we obtain a rectangle effectively comprising 12 squares. However, to obtain the posture of the square, we need to rely on the numerical triplets, which are always accompanied by the number of squares on the square.

The first Punu mask focuses on the square, while the second refers to the rectangle. The first thing to note is that both are quadrilaterals, and the first offers a more reduced view than the other. The complementary nature of these geometric figures is explained by the notion of a numerical triplet. The first is inscribed within the second, based on another geometric figure that we are about to discover. There is no such thing as chance, because every existence is subject to a previous programming whose origin is not necessarily known by the newly created element, or by the observer. One must disappear for the other to appear, yet this other is in no way different from the second because it carries within it its seeds. The father is present in the bosom of the son even though he predates the latter, just as through the son we can see the father, for they are one. He who has honoured the father will also end up honouring the son out of love for his progenitor. "The African masks show that life is cyclical around a single energy that undergoes innumerable mutations, yet plurality is not to be seen as unity. The world seeks light and not its source, which is a monumental error to be pitied". Mathematics is used as a practical and highly demonstrative language, allowing a more fluid explanation of universal realities in the African context.

5-) Triplets and the triangle

As far as trilateral figures are concerned, the arithmetical triplets obviously serve as the basis for their construction. In other words, they determine both the length of the sides and the axis of symmetry passing through the centre of the sides. Naturally, we need to define once again their different mathematical formulae: $T(n)=n+1+n$ and $T(n)=n+(n+1)+n$, whose corresponding numerical application may be, $T(3) = 3+1+3$ and $T(3) = 3+(1+3)+3=3+4+3$. The first triplet defines three aligned points, undoubtedly a situation of central symmetry. The second triplet refers to three aligned points in an axis of symmetry situation. The number 4 in the centre of the symmetrical boundary between 3 and 3 is a clear indication of this. In a triagular situation, the first defines an isolated triangle without the presence of a particular straight line. The second shows a meteoric rise at the centre and above the limits of the angle in height. La collection inter africaine de mathematiques 4eme reports that:

"The svmK'lrie axis of an isosceles triangle is at once the mëdiatrix of its base, the bisector, the height and the mëdiane passing through its principal vertex."

The arrangement of the cordophones and the horse within the traditional Mvet Ekang instrument has made it possible to define these different triplets, which fit perfectly with the configurations of the triangle. While the handwritten document does not make it possible to establish a logical relationship between the geometric figure and the arithmetical interpretation, the sculpture does. It is now possible to know the measurements of each triangle contained within the Mvet Ekang and Ngombi instruments, for example. If the base is 7cm long, then the axis of symmetry passes through the number 4, dividing it into two equal lengths of 3cm. The number 4 therefore symbolises the bridge through which the cordophones pass, 3 on each side, or simply 3 cordophones if the left and right portions are counted as a single unit. In short, this is an isosceles triangle with a base measuring 7cm and two sides each measuring 3cm in length.

If we look closely at the first and second Punu masks, we can see that both the triangle and the rectangle are represented three (3) times on them, on the forehead and on the sides. The trilateral figures on the left and right sides symbolise the base, while the one on the forehead defines the height. The use of the triangle in the descriptive definition of the Punu mask is therefore only a Hazard :

"The ënigmatic face of the mask is tegerement triangular."

The geometric figure contains countless symbolisms relating to the unity of earthly entities with those living in the heavens, which once again gives the mask an enigmatic character. It is not a fixed image, but a window leading to universal facts. No one can question the fact that we are travelling within this ancestral jewel, which opens the doors to an understanding of the universe. However, it would be interesting to look at the relationship between the triangle and the rectangle under the impetus of the arithmetical triplets.

6-) The triangle within the rectangle

In the introduction to this observation on the masterpiece that is the Punu mask, it was said that the vertical square is inscribed within a virtual rectangle, and it is this quadrilateral that we are going to highlight. However, the triangle is also inscribed within it, and it is this triangle that gives rise to the square and the crossed rectangles.

The ascent triplet $T(n)=n+(1+n)+n$ is used to create this quadrilateral. The sum of $n+1+n$ symbolises its length, while the sum $(1+n)$ refers to its width. In numerical application, we would have a length of $5+1+5=11$ and a width of 6cm, i.e. $5+6+5$. The triangle pointing towards the upper level passes through the right angles of the rectangle, or at the lower symmetrical points, and points to the upper centre (6). The triangle pointing downwards, on the other hand, starts at the upper right angles or at the upper symmetrical points and points towards the lower centre (6).

The intersecting triangles form not only a vertical square at their centre, but also two intersecting rectangles in the image of straight lines. A visual perception of the above is of the utmost importance:

Fig: triangles inscribed in a rectangle

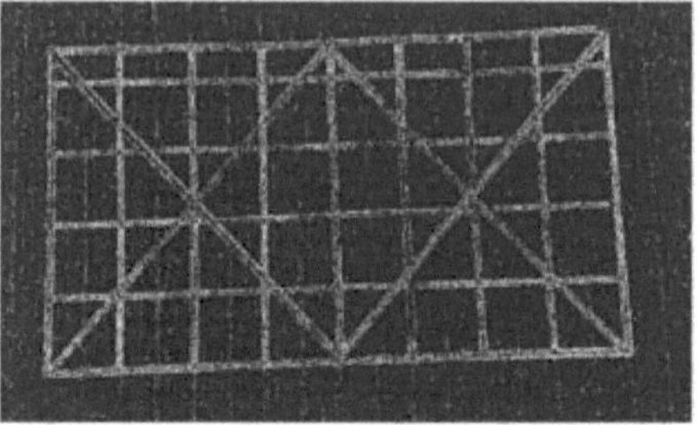

Source: NNANG EBANE Sosthene Tresor

The numerical triplet of the ascension that was used to create this compilation of geometric figures is 4+5+4, so the width is naturally 4+1+4. It's a rectangle measuring 9cm long and 5cm wide, because 4+1+4=9 and 1+4=5. The central value of the triplet symbolises the width of the rectangle, and the symmetrical axis of the isosceles triangle.

There is not the slightest doubt that the intersecting triangles form not only a square within them, but also two intersecting rectangles.

The first Punu mask focuses on the square as the central element in this compilation of geometric figures. It is the very heart of the representation, being present in both the trilateral figures and the internal rectangles. The square acts as a harmoniser, unifying opposites.

Although the mask shows a feminine face, it also reveals masculinity, because it is the woman who gives life, but this life is only made possible by her attachment to the man. She is thus the centre that unites the child with its parent. By studying one element of a given whole, it is possible to understand not only the object of study, but also the environment in which it evolves.

The second Punu mask focuses on the rectangle, which is also both external and internal to this compilation of geometric figures. It is important to note that the rectangle occupies different postures, both natural and secant. It acts as an envelope for the square and the rectangle, while at the same time enveloping itself. Several moral values can obviously be derived from this observation:

✓It's important to take care of your fellow human beings, but you also need to take time to satisfy yourself.

✓There are times in life when you have to know when to retreat.

✓life is made up of ups and downs.

✓We need to reflect from the outside who we are on the inside

✓And, more.

In these phrases, we see a dynamism that is both increasing and decreasing, where geometric shapes and figures, and numbers, necessarily convey messages relating to everyday events.

It is however vital to be interested in the exclusion of the external rectangle giving place to the hexagram.

7-) The hexagram

We need to move on from this mathematical analysis based on the Punu mask to focus on a well-known religious symbol, the Star of David. Nowadays, it is generally known as the symbol par excellence of the Jewish religion, but the mathematical approach is the only one to be highlighted here. From this point of view, the mythical symbol is presented in a variety of forms, as this passage rightly justifies:

"From a purely eslheliac point of view, it corresponds to what we call a hexagram. This gëomëtrical shape could be dëcribed by the superposition of two triangles, one pointing down and the other up."

This geometric definition of the glorious symbol fits in perfectly with the mathematical discovery of the Punu mask, as they share the same unique shape. It is appropriate to speak of the same signature or geometric inscription, which certainly creates an unsuspected parallel between them. However, it is fair to point out that the hexagram of our analytical approach comprises isolated triangles, whereas the trilateral figures known as equilateral (Star of David) make it up. What's more, the points of the triangles extend beyond their opposite bases, i.e. the virtual rectangle. An image perception of the famous star is not at all abused for a clearer understanding:

Fig: The Star of David

Sources: https://www.laportedubonheur.com>history-and-significance...

As with the discrepancies listed above, the same geometric figures inscribed within the rectangle are exactly the same within David's shield. The square is always located at the heart of equilateral triangles and crossed rectangles; in short, only the nature of the triangles changes. There is no doubt that this symbol is a common heritage of the Hebrew and Punu peoples. It would therefore be difficult to say exactly which of these peoples possessed the symbol before the other.

The Punu mask is a hexagram carved on wood and resting on a female face, no doubt to refer to her ability to build a family while at the same time being able to destroy it, based on the symbolism of opposing triangles. In fact, once she has

83

become pregnant, she alone can decide whether to terminate the pregnancy or let it run its full course. In this way, the choice of life is always revealed as a question of the utmost importance, one that humanity must face throughout its existence. Free will is a divine gift.

8-) The rectangular light panel

The notion of triplicity generally refers to a single element permeated by three different forms; isn't it right to say that perpetual motion has three entities: the body, the soul and the spirit? The idea of a luminous painting thus relates to the triple triangular variations observed within a squared rectangle, according to the value of a given number taken as the construction scale. The crossed triangles thus point to different arithmetical values depending on the number taken as the construction scale. We just need to reconsider the hexagram inscribed within a squared rectangle in order to discover it in a practical way. The following is a perfect illustration of this fact:

Fig: the light panel

Source: NNANG EBANE Sosthene Tresor

The number 17 is taken for the construction scale under the triplets 8+1+8 and 8+9+8. There is another trick to defining the degree of stagnation of the triangle knowing its regression triplet. You need to subtract two (2) units from the sum of the triplet. We have: 8+1+8=17 and 17-2=15, so the triangles each point to the value 15. The evolutionary curve of the triangular stagnation is a function of the value of its symmetrical bound, so the number 15 appears 8 times at each vertex. The triangles point to the value 120 according to the tripartite framing of the triangle, because the numerical condensation is 2+4+6+8+10+12+14+(15x8)+14+12+10+8+6+4+2 equals 56+120+56.

The triangles in this luminous painting are effectively equilateral, like those in the Star of David, as each side measures 17cm. The shield of David necessarily points to arithmetical values, as we believe it is taken from a luminous painting. Given that the length of the symbol taken as an illustration above is around 5 cm, and that the regression triplet is 2+1+2, we can work out its stagnant value, i.e. 5-2=3. The tile is 3 cm long and comprises 4 tiles. The luminous table can be used to reconstruct the geometric antecedent of a triangle, a square or a rectangle, based on its length or the number of squares that make it up, by assigning it a tripartite frame relative to its arithmetic evolution. The symbol thus belongs to a series of geometrical figures, from which it was extracted for reasons aimed at concealing a certain know-how from a certain category of people, as it is said that one necessarily returns from an unknown place before this world. The geometrist, like God, has the ability to reintegrate an element into his universe of essence. From this perspective, geometry is defined as a reconstructive awareness of a given element's membership of a family group. It is the departure from unity towards plurality.

The light chart also makes it possible to define with certainty the configuration of the triangle that a Mvet Ekang instrument comprises, according to the number of stringophones it contains. It should be remembered that the bridge is always located in the centre of the base from which the cordophones are actually clamped, so that the number on the left is the same as the number on the right, and it serves not only as the apex but also as a tunnel for the latter. In practical terms, the present triangle is a geometrical take on the Mvet Ekang made up of 8 cordophones.

9-) The square light panel

The luminous array under the frame of the square is not constructed on the basis of arithmetical triplets, but rather by taking as its length any natural number other than zero (0). However, the notion of a triplet also appears in relation to the diagonals of the latter, which play the role of axis of symmetry and triagular base. It should be pointed out that two geometric shapes make up the luminous array, namely the triangle, which appears four (4) times, and the square, which serves as the basic geometric shape. Let's now take a look at the following light table:

Fig: un carré quadrillé

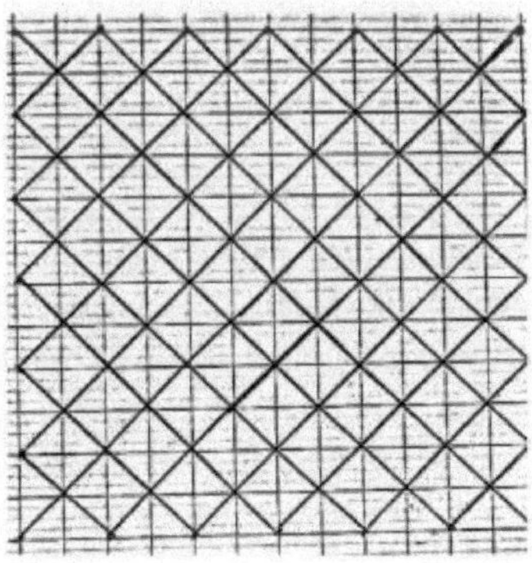

Source: N'NANG EBANE Sosthène Trésor

The numerical summary of this table is as follows: 2+4+6+8+10+10+8+6+4+2=60, so the number 20 represents the degree of stagnation. This stagnation within the tile is concentrated only on the diagonals, forming two columns of 10 tiles each. The diagonal as an axis of symmetry makes it possible to reproduce the

85

image to which it forms the base, or infinity, on the opposite side to that image. So the triplet 20+20+20 effectively shows that the number 20 in the centre symbolises the diagonal, while the others define the opposite images.

The diagonals are also the two top sides of each triangle in relation to its base. They each have 15 squares and are each 7cm long.

10-) Diagonals and triangular stagnation

It was shown above that the rectangle results from the modification of the square, from a miniature perception to the so-called grandiose one, based on the numerical triplets. In this new analysis, we need to illustrate this metamorphosis on the basis of a purely arithmetical relationship. The data in the rectangle reflects the diagonometric readings of the square. Let us consider the following with particular attention:

Diagonals:T(n) = n+1+n+T(n) = n+(n+1)+n = Stg(T(M)). The sum of the diagonometric triplets of a given number is equal to the sum of the numerical condensation of this number, in a luminous array. In a square of length 7 cm, the value of the diagonals by their triplets varies from 5+1+5=11 to 5+6+5=16, hence 11+16=27. The triangular numerical condensation revealed by light table 7 is respectively: 2+4+5+5+4+2=27. In other words, it is the numerical evolution of the diagonals of the square that is revealed in the light table.

The number 8:

We have: 6+1+6=13 and 6+7+6=19, hence 13+19=32

Numerical condensation: 2+4+5+5+5+4+2=32 or 5+27=32

The number 9

7+1+7=15 and 7+8+7=22, so 15+22=37

This mathematical formula relating to the addition of a descending triplet with the so-called ascending triplet does not fit with the sum of the numerical condense of the number 9 in a luminous array starting from the diagonals of the square. It would therefore make sense to write: [(n+1+n)*2]+[n(n+1)+n] equals [(7+1+7)x2]+[(7+(7+1)+7)] = 15*2+(7+8+7)=30+22 = 52. This difference is undoubtedly due to the fact that the numbers 7 and 8 are not perfect triplets, i.e. 3+1+3=7 and 3+2+3=8, whereas the number 9 is a perfect triplet, 3+3+3=9.

In the final analysis, it is the reading of the diagonals of the square by addition that delivers the luminous picture, which is in reality a rectangle in the form of triangular numerical condensations. The diagonal is no more and no less than the base of the triangle within the square, which in reality comprises four trilateral figures. We thus move from a configuration of squares and triangles to another known as a triplet: rectangle, vertical square and crossed triangles.

11-) Natural tile and vertical tile

Having established the relationship between the length and internal value of the previous vertical square, it's now the turn of the squared square to reveal its algebraic organisation. An image of this is of vital importance.

The square above is 7 cm long. It contains two internal values that suggest the idea of undulation, as the first is one unit shorter than the second, which is the longer value. They are respectively less than the length of two (2), and one (1), which refers to numbers 5 and 6. To calculate the total number of tiles, multiply VI1 by VI2 and VI2 by VI1, then add them together. Vtc= (VI1xVI2)+(VI2xVI1) equals (5x6)+(6x5)=30+30=60. To determine the degree of diagonal stagnation, Dgst=(VI1+VI2)-1 equals (5+6)- 1=11-1=10. In total, we count 4x10=40 as the overall value of stagnation. It should be noted in passing that the diagonal includes a numerical value less than the length of the square of a small unit, i.e. 6 and 7. Adding them together corresponds exactly to the number of parallel line segments in fractal dimension that this square comprises, i.e. 11 in one direction, and 44 in the 4 directions.

It is therefore possible to know the numbers associated with each digit used as the length of a given square:

The number 8:

VI1:6, VI2:7, Vtc:84, Dgst:12 and 48, VI1+V2:13

The number 9:

VI1:7, VI2:8, Vtc:84, Dgst:14 and 56, VI1+ VI2:15

From the above, it can be seen that the vertical squared tile has a stagnant internal numerical value (---) which is higher than the square. On the other hand, the normal grid tile has two internal values describing an undulation (   )

continues. Stagnation and mobility are once again expressed within these squares. The waters of an ocean are both calm and agitated, according to the specific seasons of tide on the one hand and flood on the other, a clear transposition of death and life into a common, regular perception.

Diagonometric triples are numerical triples of both ascension and regression, where the same elements present in the ordinary tile are also the same in the vertical tile.

12-) Triangular numbers and stagnation

Both the moon and the sun belong to a single sky, despite the fact that each has its own hour of glory, or simply its own time to illuminate it. In the same vein, even and odd numbers belong to the domain of arithmetic, but each genre has its own speciality. It is therefore appropriate to study their influence within a trilateral figure. This analysis will begin with the odd numbers.

Example 1: the number 7

It has been shown above that the 7cm long tile has four (4) triangles made up of 15 tiles each. In addition, according to the triangular numerical condensation of the luminous array as follows: 2+4+5+5+4+2=27, the stagnation is 15°.

Example 2: the number 9

The square is 9 cm long and contains triangles with 28 squares each. The data from the rectangular light box also reveals a stagnation identical to this number: 2+4+6+7+7+7+7+6+4+2. The number 7 appears 4 times, which refers to the number 28.

In the case of odd numbers, the triangles point to the same value that makes up the number of squares they contain. They thus reflect their inner values from the outside. The arithmetical values of the trilateral figures are exactly the same in both the luminous square and the luminous rectangle. This is a very instructive moral lesson: let's expose who we really are on the inside. The triangle is most often known as a figure of size, but it is exactly these different constructions that demonstrate it even better. Under the prism of arithmetic triplets, odd numbers are defined as the soul of trilateral figures, because they are conspicuously absent from these hollow shapes. The unspoken is something to be sought by the avid observer. However, it is also important to consider the configuration of even numbers. The idea here is to establish a correspondence between the number of squares in a triangle and its degree of stagnation. So it's an internal and external relationship that comes to the fore. The question arises as to how to determine the internal value of a triangle knowing its degree of stagnation, or how to determine the degree of stagnation knowing the internal value? The degree of stagnation we're talking about relates to the number that can be seen in the numerical condensation of the light panel. This particular configuration focuses exclusively on even numbers.

Even numbers have a different configuration to the one above. This is because the internal value of the triangle differs by a small unit from the degree of triangular stagnation in a corresponding light picture. It is important to understand that even numbers do not have a centre of equilibrium equal to 1. Dgst (degree of stagnation)=Vt (triangular value)-1.

Consider the following figures:

Example 1: the number 4

Triangular value: 3

Degree of stagnation: 2

Example 2: the number 6

Triangular value 10

Degree of stagnation: 9

Example 3: the number 8

Triangular value: 21

Degree of stagnation: 20

The last example counts four (4) times the number five (5) repeated consecutively. We have: 2+4+5+5+5+4+2=32. The number 5 symbolises the degree of stagnation of the triangle, from which each trilateral figure necessarily points to a precise arithmetical value according to such and such a measure taken for the length of its sides. Both the number 8 and the number 7 have the same stagnant value of 5.

For odd numbers: the number of squares counted in a triangle is equal to its degree of stagnation within the luminous array, so Vt=Dg.

As far as even numbers are concerned, the numerical value of a triangle is greater than that of its degree of stagnation in a small-unit luminous array, so Dg=Vt-1.

13-) Diagonometric triplet and triangular stagnation

The relationship between the luminous square and the luminous rectangle is based on diagonometric triples and triangular stagnation. The degree of stagnation of a given triangle can be determined from the diagonometric triplet of the ascension of the square.

The degree of stagnation is simply the downward revision of the central figure or number of the ascension triplet. We have 5+6+5=16 from which we pose 5+(6-1)+5=5+5=15, designating the right angular stagnation of the triangle. The triangle points to a value of 15° according to the chords of the number 7. So the increasing triplet 5+6+5 defines the evolution of the diagonals of a square whose side is 7 cm long.

The appropriate mathematical formula is defined as follows: Stg(n)=n-1+(n-1)+n

Example 1: the number 5

stg(4)=4-1+(4-1)+4-1=3+3+3=3 x 3=9

The triangles point to the upper and lower 9°.

Example 2: the number 7

stg(6)=6-1+(6-1)+6-1=5+5+5=3 x 5=15.

The triangles point at the upper and lower 15°.

Example 3: the number 9

Stg(8)= 7+[8-1+(8-1)+8-1]= 7+(7+7+7)=7+(3x7)=7+21=28

The triangles point to the upper and lower 28°.

It should always be borne in mind that in a square, the main or central diagonal is symbolised by an arithmetic value that is always less than the length of the side of the square of a small unit. So, the stagnant values denoted Stg(n) designate the diagonals of each number.

However, it is important to emphasise in passing that the diagonometric triplet only provides information about the first three digits or numbers making up the stagnant triangular values of a given digit or number. It is therefore strictly recommended that you always refer to the luminous table to find out how many times the stagnant digit or number appears. The first number 7 in the third example is an addition from the rectangular lightbox to define the real arithmetical value of stagnation of the number 9.

IV-) SONGO

Source :https://www.wikipedia.org>wiki>le-songo

A
special thought for
Mr MEYE M'EYEGUE Paul Marie
(Man of culture)

1-) arithmetic triplets and parallelism

Triplicity seems to be one of the notions common to all the objects of study recorded in the present analytical approach, via observation with the naked eye. The backbone of the Songo, or the calculation game of the sowing family, is suspected of being drawn from a geometrical figure that gradually tends to emerge. However, it would undoubtedly be wise first to dwell on the second perception of the number 7, in order to better bring out the other face of the object of study, the Songo:

"It is played by two players and is dividedë into 2 territories. Each side consists of 7 squares.

In fact, the number is one of a series of arithmetical triples whose decomposition reveals a perfect balance between the values, with the extreme right value set against the extreme left, in relation to a central value that is always equal to one (1). The regression triplet is noted: $T(n)=n+1+n$ equals $T(3)=3+1+3=7$. This decomposition reveals geometrical notions such as central symmetry, orthogonal symmetry, perpendicularity, trilateral figures (triangles), etc,. The transition from the monopartite vision of the figure to another, called tripartite, shows that multiplicity is present within unity, and that unity is also manifest in diversity. The Songo squares are multiple from an accounting point of view, but they are no longer object-based, like a global perception. Only odd numbers have this configuration. Even numbers, on the other hand, have a symmetrical unit that is different from 1. Triplicity is an expanded or compartmentalised perception of unity in three dimensions. The notion of arithmetical symmetry is evoked in exactly this way. The fourth (4th) square of the game is therefore the centre of symmetry of the squares. By translating this reality into triangles, which symbolise the symmetrical unit 1, the present symbol emerges:

Fig: the light panel

91

The vertex of each triangle points to the fourth square in the column of 7 squares opposite it. The points link the top number 4 to the bottom number 4. The game also offers a parallel correspondence of numbers from 1 to 7, i.e. 1234567 (higher) against 1234567 (lower), like the Ngombi cordophones or the Sacred Harp.

In general terms, the crossed triangles represent the balance of forces between the heavens and the earth, which can also be translated as the unity of masculinity (triangle pointing to the heavens) and femininity (triangle with the apex pointing to the earth), like the permanent cycle of the universe (Source triangle and its symbolism).

This parallelism here refers to the players standing face to face on the game board, able to manipulate the seeds by moving them from one square to another. The action of one implies a simultaneous reaction from the other, with each action having a consequence proportional to its magnitude. Rain falls from the heavens, and plants grow from the earth as a direct consequence. If rain expresses regression, plants define ascension, hence the perfect balance of nature. The players are the personified transposition of the organisation of the universe in a cyclical artistic take.

The players, standing in their respective camps, carefully observe the seed throwing technique used by their opponent to win the game. The geometric symbol in the shape of an open eye, as an organ of light, naturally refers to vigilance, which must be connected to intelligence in order to make an informed decision.

remarkable part of the game. An attentive eye and a preventive memory must be the qualities of a good Songo player. In reality, they each start at the base of the triangle, but as the game progresses, the player who loses the most seeds is seen to be lower than the other. The player's height is a function of the number of seeds he practically possesses, but the implicit reality places him at the apex of the triangle. Since the goal to be reached is a projection of thought, the journey towards that goal is a fact that eventually becomes concrete. In this way, we are practising the work of creation, moving from thought (height) to action (base).

Reflection presupposes an exact recognition of the world of intelligences that humanity must visit when the need arises. However, as not everyone is able to distinguish it, the game becomes a window through which we can enter it easily and unconsciously. How do you get the better of your opponent, if you know what to do when the going gets tough? Winning or losing is not an end in itself, it's knowing how to be unflappable. The seasons come and go, but time remains the same. Man undergoes multiple transformations through time, which always ends up initialising the elements living within him. His achievements are shaped by time. Each new part is the work of initialising the elements undergoing its action. Patience is the reason for inexplicable success.

2-) Square and triagular corrugation

The parallelism expressed by the opposing rows of 7 squares each leads us to

92

question the relationship between the number 7 and the number 5. At the outset, the number 7 was revealed as belonging to the family of arithmetical triplets, and offering a perception that is both restricted and enlarged or compartmentalised, but it was not implied that it occupies third place in this classification. It must be understood that the number 5, on the other hand, occupies second place, and therefore follows it. Note that T(n)=n+1+n equals T (2)=2+1+2=5, so n is equal to the value of the symmetrical boundary, and to the rank occupied by the corresponding digit. How does the number 5 integrate the number 7 as an internal value? It would no doubt be easy to say that the latter is superior to the former, but this answer will not be sufficient to show why each square has 5 seeds on one side? And why are they hollow, like the gourds of the Mvet Ekang? We should rightly consider a square 7 cm long, with internal values varying between 5 and 6. The first (5) indicates its lowest level, while the second (6) indicates its highest point. The 7 cm long squares show an undulatory arithmetical variation from 5 to 6, and from 6 to 5 to infinity. The seeds placed in the squares are the same as those collected by the players, which means that the players grow and shrink throughout the game, which corresponds perfectly to this undulatory variation. The number of squares represents the number 7, the seeds symbolise the number 5, and the number 6 refers to the wooden blocks separating the squares. The waves are created by the player's arm moving from one square to another.

The undulating variation through the numbers 5 and 6 demonstrates to some extent that life is made up of ups and downs. The players are fully aware that they cannot win everything in every game they play, and that defeat invites them to remobilise in order to come back stronger. The number 5, symbol of the fall, is thus associated with the number 6, symbol of growth, while the hand refers to the energy that enables not only the fecundity of the earth, but also water as a reference to fertility. It is said that the Songo is part of the sowing games, a fact that is not only idealistic, but also practical.

Geometry is thus linked to human activity, as a kind of descriptive writing, but not perceptible to the naked eye. The idea would be that humanity, through its actions, is writing its history in a purely geometric dimension. A man standing is similar to an ascending curve, while the fact that he is seated indicates a descending one. A non-zero natural number in the layout of a square is delimited by its ascending and descending degrees. They therefore define here the bipartite anteriority of this number. A father and mother are anterior to their son, but the latter is an added value of this union. In him lies the portion of each of the parents, and his personality.

3-) Internal triangular components

The notion of triplicity is always a weighty one, even in a quadrille tile, as it incorporates triangles through the action of the diagonals. It was said above that the length of the square is 7 cm, but the trilateral figures within it are each made up of 15 squares. The number 15 divided by 3 gives the number 5. In the light of the above, it is easy to understand that the number 5 is internal to the number 7, as one designates the triangle, which is internal, while the other symbolises the external tile. In the number 15, the number 5 appears 3 times, to signify its manifest presence in the various Songo squares. The 7cm-long tile comprises a total of 4 triangles of 15 tiles.

The relationship between the square and the triangle as components of the same geometric figure, according to internal and external postures, or whether the triangle is inscribed in the square, seems to fit in with the framework of the sowing game. The seeds, of which there are 5 inside each square, are the same as those carried by the hands of the players passing from one square to another. As the harvest is an action that passes from the bottom of one square to the top, ending up at the base of another square that is opposite the first, it is a practical triangular sketch. In contrast, the squares arranged in two columns reveal the presence of the square. But which part of the square do they refer to?

Generally speaking, the relationship between the square and the triangle gives rise to a unique geometric figure, the pyramid. The quadrilateral usually refers to its base, while the trilateral refers to its height. The former is considered to be matter, while the latter refers to the work of the spirit, which gives life to it (the symbolism of the pyramid).

The name "sowing", which is used to define the Songo game, is not far-fetched, as it relates without the slightest doubt to the idea of creation, growth, greatness, etc. The hand is thus the author of this sowing and, by immediate effect, of this harvest sketched out through the game. The hand is thus the author of this sowing and, by immediate effect, of this harvest sketched out through the game. It is capable of both adding and subtracting, of giving and taking away. It is productive, which is why the fieldwork is so famous throughout Songo. Since the hand is the author of creation, which is symbolically represented by the pyramid and therefore the Songo, this architectural marvel is the work of the divine. The entire universe would therefore be under the control of an invisible hand governing its functioning. Some energies are purely terrestrial, like the square, while others are celestial, like the triangle. In the same vein, there is always a winner (triangle) and a loser (tile) in every game. In short, the Songo game is an unconditional recognition of the existence of an extraordinary intelligence, symbolised by the hand, the author of all creation, in the form of the human activity of farming.

4-) Diagonometric stagnation of the edge

Still considering the 7cm long square, the practical correspondence between the arrangement of the Songo squares and those of the diagonals will be brought to the fore in this section of the observation. The square adopts certain arithmetical properties proportional to the figure or number taken as the length of the square. So, if the length of the square is 6cm, its main diagonal corresponds to the number 5, while 4 secondary diagonals in a fractal form surround it on either side. The corresponding diagonometric triplet is therefore 4+5+4=13. As for the number 7,

94

we have: 5+6+5=16. The number 6 thus symbolises the main diagonal, while the number 2 5 or 5/5 designates the secondary diagonals on either side of it. The number 5 on the left is symmetrical to the number 5 on the right in relation to the number 6, which is their arithmetical centre of symmetry. The number of seeds in each square of the sowing set is therefore relative to the arrangement of the diagonals within the 7cm long square. It should be understood that each semi-circular block of wood separating the squares is equivalent to the number 6 as a numerical value. The squares of the sowing game would therefore be a practical transposition of the diagonals of the 7cm long square, in a triplet form.

The traditional game was designed not only to give players a sense of fulfilment, but also to illustrate that every species of nature, or quite simply every creature, can be recognised by one or more of its own codes. The number 7 has undoubtedly been chosen for particular reasons, but we realise that it cannot manifest itself without its arithmetical antecedents, i.e. without the numbers 5 and 6. In the family context, the child cannot come into the world without the joint will of the father and mother. The parents therefore represent the beginning of the child's history; the child, on the other hand, is the present moment. Songo's anteriority is therefore purely mathematical in origin, through the manifest unity of the triangle and the square. However, it is clear that the number 7 is strongly linked to the number 5, which is not the case with the number 6, which is very close to it. It is sometimes observed that the child is more attached to one of the parents, either the father or the mother. Both are odd numbers, and therefore arithmetical triplets of the same nature, which is unfortunately not the case with the number 6. All things considered, triplicity seems to define the introduction of a foreign element into a complementary binomial, leading the two to solidify their union.

5-) Rectangle and triangular stagnation

In this section of the observation, we need to go from the square to the rectangle, using a triplet perception of the length of the former. The length of the square is 7cm, and its corresponding triplet configuration is 3+1+3=7, which is used to construct the rectangle. The regression triplet is transformed into the ascent triplet by increasing its arithmetic centre of symmetry by 3 units, i.e. 3+(3+1)+3=3+4+3=10. However, the triangle within the square at the start is exactly the same within the triangle at the finish. It still consists of 15 squares. So, the triangular numerical condensation appears with the triplet perception of the number 5, i.e.: 2+4+5+5+4+2=27. The number 5, which constitutes the number of seeds in each of the Songo squares, is a triangular stagnant value of the number 7. The square triangle has 15 squares, as does the rectangle, but it offers a decomposed perception of the number 15 in three appearances of the number 5. A rectangle 7cm long and 4cm wide contains a 15-tile triangle and triangular numerical condensations in which the number 5 appears in three dimensions.

The 5 seeds in the sowing game are therefore a singular perception of the number 15, which represents the degree of triangular stagnation of the number 7. The hand

95

powering the seeds from one square to another would therefore draw a triangular sketch whose vertex is worth a stagnant value of 15. It would therefore move from 2 to 15, then from 15 to 2 on the basis of the opposite side. In view of the above, the Songo players would be performing mathematical activities through their simple gestures. The latter would therefore be evaluated from a purely geometric point of view. Human activity would therefore be a practical transposition of the science of mathematics.

6-) Quadrilateres and triangular stagnation

Discovering and understanding the sowing game from a mathematical point of view leads to determining the total number of seeds available. It would therefore be no mistake to dwell once again on the following descriptive passage:

"It is played by two players and is divided into 2 territories. Each side consists of a row of 7 squares, each of which contains 5 seeds at the start of the game."

So far, the emphasis has not yet been placed on the part of the tile corresponding to the practical framework of the game. This specificity will be the subject of a subsequent mathematical analysis. The current preoccupation is based jointly on the square on the one hand and the rectangle on the other. The quadrilaterals share a common feature relating to the numerical evolution of the triangles they incorporate. The square offers the following numerical condensation: 2+4+6+8+10+8+6+4+2=60, the stagnant value of which is 2x10=20. As for the rectangle, it has a completely different configuration: 2+4+5+5+4+2=27, whose stagnant value is 3x5=15. We naturally need to add them together, so 15+20=35. So the total number of seeds in a row of 7 squares, each containing 5 seeds, is 35. Simply multiplying 7 by 5 is not out of the question, but we're still in an analytical phase whose aim is to discover and understand the object from a mathematical point of view. The number 35 therefore refers to the unity of the stagnant values of the square and the rectangle. The total number of seeds contained in the Songo squares relates to the unity of two squares with two rectangles, i.e. each of the squares is associated with a rectangle. This could be translated as: Square1+rectangle1 (North side) and square2+rectangle 2 (South side) making up each row.

The seeds in the traditional Songo game would, in the light of the foregoing, be the sum of the joint heights of the square and the rectangle, through the actions of the triangles they incorporate. The number 70 relating to the seeds contained in the object of study, has as its geometric reference the inscription of the triangle within both the square and the rectangle. It would also be possible to multiply the length of the square by one of its diagonometric stagnations, i.e. 7x10=70. The fact that the game can be built on a surmounted support also shows, to a lesser extent, that it is the product of heights. What's more, the game is played with the hands that make up the upper limbs of the human body. A multitude of symbolisms relating to the distancing of the object in possession from its possessor are clearly made explicit through this game. The object being manipulated and the manipulator share nothing

96

in common, apart from the fact that they are oriented by it. From this point of view, we always come back to theological thinking about a supreme Being who gives life to everything he touches. The seeds live under the repetitive cycle of the players throughout the game, who sometimes attribute extraordinary powers to them through their speeches: Here's the seed that's going to knock your socks off! they sometimes shout.

7-) Stagnation of the edge in 10

The end of the game takes place according to particular configurations. The number of seeds inside the game deck must not be less than a certain number. However, what effect would this have on the geometric structure? It's worth highlighting the relationship between the number of inferior seeds and the diagonal forming the knot of the tile. It is precisely this particularity, centred around the diagonals, that has led to the use of the notion of diagonometry. In other words, an observation devoted to the different roles of diagonals in geometric figures. It has been observed that they can relate to the triangular base, the axis of symmetry, the diagonometric triplet, etc,. The number 10 will therefore be used as an example in this situation. The following explanatory passage relating to the end of the game will serve as a focal point:

"When there are fewer than 10 seeds left in the playing deck."

The diagonals are inside the tile, and form a right angle of 90° at the centre of the tile. This illustrates, in the most ordinary case, that the diagonals necessarily relate to a certain numerical value. The configuration of the squared square further amplifies this mathematical reality, as it allows us to recognise the length of a square in relation to its degree of stagnation. The number 10 is a stagnant numerical value of the 7cm long square, due to the parallel doubling of the diagonals. It can be seen that the quadrilateral has arithmetical values that are both static (the length of the square) and evolving (the diagonals). The centre or interior of the square is clearly mobile, while the exterior is not. Added to this is the fact that the same diagonals also constitute the triangles inscribed within it. As a result, the triangle is most often defined as symbolising the breath, which gives life to the matter embodied by the square (the symbolism of the pyramid). The human body can effectively be transposed here through the flesh, which is famous for giving life to the body's envelope. Returning to the endgame, the presence of less than 10 seeds on the board implies the destruction of the diagonals that make up the very soul of the quadrilateral. This number Y is, in reality, the stagnant value of a square with a length measurement of X, inevitably less than 7. The end of the game is imminent because it no longer corresponds to the numerical order of the game board.

8-) 8cm square columns

The presence of the number 8 in the analysis may seem insane to some, who believe that the numerical order of the deck is no longer taken into account at all. It was said above that the square has properties such that the number used to measure its sides incorporates a central diagonal, corresponding to a figure or number that is one less than it. In order not to lose the thread of our ideas, it is essential to have a visual perception of this square:

Fig: a squared square

Other than the diagonometric reading, the destruction of parallel and perpendicular lines gives way to points aligned in columns. In this case, the columns are each made up of 6 squares, which does not correspond at all to the number 7. The practical layout of the 14 squares of the sowing game needs to be investigated. Based on the number 8, note that its main diagonal corresponds to the number 7. The corresponding diagonometric triplet is 6+7+6=19. The undulatory variation can be read from 6 to 7, then from 7 to 6, so the corresponding symbol is 6<7>6. On the basis of this observation, we can hypothesise that the Songo squares are a transposition of two columns of 7 squares, the model for which is taken from a square measuring 8 cm in length. Let's now look at the practical view of the columns of the quadrilateral:

Source: NNANG EBANE Sosthene Tresor

Four points are considered to be the equivalent of a Songo square. From this point of view, there are 7 columns made up of 7 squares. Given that parallelism is a striking reality on the game board, it can be argued that this does indeed fit with the above diagram. The recurrence of the number 7 within this 8cm long square reinforces this fact once again. The number 7 cannot be expressed within the tile whose sides it is taken to be, but within another whose measurement is one unit greater, i.e. 8cm.

In this respect, numbers are structured in such a way that the value of the smallest number is an integral part of that of the higher number. They evolve while retaining part of their antecedents, a practical transposition of the human and animal family. The child is new in relation to its parents, and includes within itself each of their parts. He is therefore in no way superior to them, because they gave him life. However, the superiority that tends to be accorded to higher numbers takes little account of the order of appearance. It is formally recommended to respect one's elders, but financial reality is the practical reversal of this morality. A banknote of a thousand CFA francs is worth more than one of five hundred, which is why it is easy to understand why God is excluded from human activity. If elders are to be respected by the young, then a five hundred franc note must be worth more than a thousand CFA francs. By simple analogical deduction, the number 7 gives rise to the number 8, so genealogy is a practical mathematical reality.

9-) Rectangle and triangular stagnation

The total number of seeds contained in the play apron has been calculated above, according to the square and rectangular stagnation peculiar to the number 7. However, the aim here is to highlight the complementarity between this number and the number 8, in terms of the total number of seeds. After demonstrating that the number 7 is indeed present in the 8cm square, we need to present the triangular stagnation degrees:

The numerical sum of the number 7 is: 2+4+5+5+5+4+2=6+15+6=27

The triangular numerical condensation of the number 8 can be summarised as: 2+4+5+5+ 5+4+2=6+20+6=32

In view of the above, the numbers 7 and 8 share the same degree of numerical stagnation, which is 5, despite the fact that the latter reveals a triangle with two vertices. This is quite natural, as even numbers do not have a centre of arithmetic symmetry equal to 1. The corresponding numerical triplet is noted: $T(n)=n+2+n$ equals $T(3)=3+2+3=8$ and $T(n)=n+(n+2)+n$ equals $T(3)=3+(2+3)+3=3+5+3=11$. The ascension triplet does indeed reveal the number 5 as the centre of arithmetic symmetry, but the symmetric limits remain the same. So we need to consider only the stagnant values, i.e. the number 15 and the number 20. By adding up these results, we obtain the number 35, corresponding to the number of seeds contained in a row of 7 squares, each consisting of 5 seeds. The 70 seeds therefore refer to the double triangular stagnation pairs of numbers 7 and 8. To put it more explicitly: $(15+20)x2=2x35=70$ or $(2x15)+(2x20)=30+40=70$.

If we look at the number 5, we see that it appears exactly 3 times in relation to 7, and 4 times in relation to 8. The number of times the number 5 appears in the two stagnations is 7. We are thus witnessing a constant recurrence of this number, what can we say about its numerical condensation obtained by adding the numbers together: 3+4+5+7+8=27? Also, the addition of the number 7 with the number 8 results in the number 15, corresponding to the triangular stagnation of the first number. Everything seems to show that this number has something special about it, because when it is added to others in this way, the results obtained are relative to its own diverse perceptions. The number 8 tends to disappear to make way for the number 7, and instead of progressing in relation to the former, we regress in relation to the results obtained in accordance with the latter. This number appears as the original source of light, but the number 8 is only one of the light rays of the latter, all the more so as it is only one small unit higher. It is therefore a logical continuation of this.

The theological thought that follows from this mathematical analysis is that multitude is simply a reflection of unity. In other words, the universe was created from a single energy, which simply multiplied to fill it in its entirety. The 70 seeds in the sowing game are the fruit of the number 7 as a singular unit. It is therefore an exhaustive view of this number, which can be dissociated, either 7+10=70 or 7x10=70. Unity and multiplicity express abundance, but subtraction, like division, expresses the recognition of the original being or the source energy. We must rightly give credit to genealogy, starting with the new generation and going back to our ancestors, who certainly communicated with God by word of mouth. Addition and multiplication are the work of both the present and the future, while subtraction and division relate to the past. Mathematics defines the passage of time, the ability to reconcile the past, the present and the future, through the manipulation of elements beyond the manipulator's vision.

10-) Square diagonals 9cm

The columns of 7 squares were presented above within the 8 cm long tile, as a possible hypothesis for their extraction from the latter. However, they did not present any practical cut-outs that were really in keeping with the framework of the traditional Songo game. They are therefore an underground view of the 8cm long tile. To achieve this, it would be possible to examine the 9cm long tile. Remember that the numbers 7 and 9 are arithmetical triplets, the third (3) and fourth (4) respectively. The addition of the postures always leads to the number 7. Let's take a close look at the square below:

Fig: the 9 cm long square

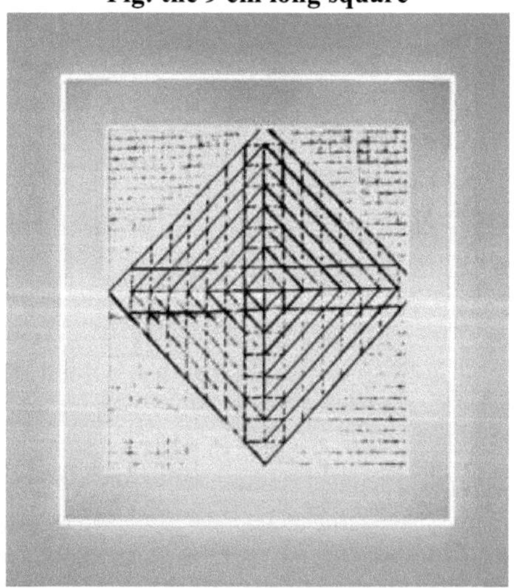

Source: N'NANG EBANE Sosthene Tresor

If you look closely at the image above, you can see that this time the diamonds are arranged according to the configuration of the Songo squares. There are in fact 4 columns, each made up of 14 squares, the sum of which is 28 in binary and 112 in quadruple. To get a practical idea of this, we need to consider the tiles running from the centre to the right angles. We will see that there are 14 tiles, arranged in 2 columns of 7 tiles.

This geometric construction reveals the arrangement of the diagonals into four parts, each made up of 14 squares, which are certainly taken to be the squares of the sowing game. As such, it is taken from the diagonal cross visible at the base of the pyramid, caught between the fractal evolutions of the squares. The squares in the staircase posture reflect a mobility that is not obvious but is perceptible to the naked eye, whereas the diagonals taken to represent the squares are immobile. Returning to the game, the squares are without a shadow of a doubt static, whereas the seeds, which both grow and shrink, express mobility. The multiplicity of seeds is therefore a practical and perpetual transposition of the squares evolving on either side of the diagonals. The winner possesses more seeds than the loser during a game, but both taken together constitute the boundary limits of both increasing and decreasing revolution of the seeds.

Let's now consider the position of the players in relation to the board. The sowing game is located in the centre of the opponents, who naturally face each other, the whole depicting the image of a cross or that of diagonals intersecting at their point of intersection. This posture places them at the very centre of the pyramid, as shown in the image above. It's important to understand that even the players blend in with the diagonals to a certain extent, to get an image that's relatively consistent with the geometric figure. In this configuration, the players are considered to be unanimous. They then symbolise a second game board, arranged in the opposite direction. The letter X in the alphabet can be seen as an image transposition of the above.

Geometric shapes are naturally age-old, but the mobility we attribute to them stems from the human desire to be able to identify with them. Representation thus becomes a medium for depositing human identity. The codes used must not be analysed from a singular perception, but rather from a general one, in order to grasp the value of geometry. The transition from 'figurative' geometry to 'animated' geometry necessarily integrates certain life-giving components within the elements to be brought to life. The traditional Songo game obeys this existential logic perfectly, through the players animating the seeds within the game board.

11-) Rectangle and triangular stagnation

It's important to understand that the triangular stagnation within a luminous array only concerns odd numbers. They are used as a construction scale and also reveal a triangular stagnant value. The even numbers, in triplet form as a construction scale, also reveal an odd number at the vertex of the triangle. However, it's the odd numbers that take pride of place.

The triplet of regression of the number 9 is 4+1+4=9 and the triplet of ascension is

4+5+4=13.

Combined within the luminous rectangle, they produce the following triangular numerical condensation: 2+4+6+7+7+7+6+4+2=52.

The triangular stagnation of the number 9 within a luminous rectangle reveals the number 7777, as a stagnant value. Remembering that the diagonals of the square are 9cm long, they are divided into 2 columns of 14 squares, the sum of which is 28. This means that one (1) column of 14 squares or two (2) rows of 7 squares each are transposed onto the game board.

The number 28, symbolising the diagonals of the 9cm square, is also an integral part of the number 7. To find this out, break the number down into 7 units in ascending order, then add them together. We count: 1 2 3 4 5 6 7, then calculate the sum: 1+2+3+4+5+6+7=28. In reality, this ascending order refers to the side of a triangle going from its base to the top, and the descending order leaves the top for the base, i.e. 7+6+5+4+3+2+1=28. Let's get back to the game:

"The game mechanism consists of picking up all the seeds from a square in your territory, depositing them one by one without jumping a square, from right to left in your camp, then from left to right in the opposing camp, until the seeds run out."

The mechanism effectively requires seeds to be deposited from the right to the left, then from the left to the right, as described above. It should be noted that the left is cited successively, while the right tends to be distanced from the other, hence the formation of the triangle: right, left, right. Let's go back to our wave values for the number 7 appearing on the square. The number 5 appears 6 times, the number 6 5 times. So to find the undulatory sum, we ask: (5*6)+(6"5)=30+30=60. It is possible to destroy these pairs made up of 4 elements to get just 3 elements, i.e.: (5x6)+(6x5)=5x6x5=150. You need to understand that the whole analysis consists of starting from the square (4) to the triangle (3), from which the number 7 emerges as the numerical symbol of the pyramid. The number 3 is contained in the number 4, although it is an independent value. Similarly, the triangle is a component of the square, even though it is a trilateral figure. In other words, the presence of one automatically implies the presence of the other. It is from this perspective that the pyramid implies the addition of opposites, even though it is not an addition in the stereotypical sense of the term, because it reveals the possibility of assimilation. The square can become a triangle, the triangle can become a square, then the square and the triangle are one and the same. Songo is not only a game, but also a real transposition of the unity of the triangle with the square, in the staging of life-giving beings (players) with beings to be life-giving (the seeds and the playing apron). Songo is no more and no less than a pyramid in perpetual evolution, or an animated pyramid, whose most striking but hardly recognisable everyday symbol remains the diagonals at its base, formed by the position of the playing board between the players.

12-) The 345 triplet and the 789 triplet

The number 3 is the boundary of the number 7, and the number of times its triangular stagnation value appears (555=3x5=15).

The number 3 is the boundary of the number 8, not its stagnant triangular value, still less its frequency of appearance.

The number 3 is contained threefold within the number 9, but does not constitute its triangular stagnant value.

The number 4 is the arithmetical centre of symmetry of 7, and its internal value (3+4=7).

The number 4 is doubled within the number 8, and constitutes its arithmetical centre of symmetry.

The number 4 is the symmetrical boundary of the number 9, not its stagnant triangular value, but its appearing number (7777=4x7=28).

The number 5 is not only an undulatory numerical value of the number 7, but also its triangular stagnant value, at a frequency of 3.

The number 5 is not a stagnant numerical value of the number 8, but constitutes its triangular degree of stagnation, under a frequency of 4.

The number 5 is the arithmetical centre of symmetry of the number 9, but is not its triangular stagnant value.

The number 7 is a triangular wave value of the number 8.

The number 7 is a triangular wave value of the number 9, and constitutes its degree of stagnation.

The number 8 is an undulating value of the number 9, but is not its stagnant value.

The number 9 is a value encompassing the numbers 7 and 8.

The sum of the triplet 345 is 12, i.e. 3+4+5=12

The sum of the triplet 789 is 24, i.e. 7+8+9=24.

In the light of the above, the first triplet constitutes half of the second, while the latter comprises its double. It is essential to understand from this that the notion of fractal is not only perceptible through geometric figures, but also through numerical activity. The least important value is always located above the most valuable one. There is an arithmetical progression from 12 to its double 24, and a possible regression from 24 to 12. The squares do indeed evolve from the base to the top in ascending order, and also from the top to the base in descending order. In other words, geometric shapes are expressed in arithmetical terms, while at the same time geometry reveals numerical activity. Geometry and algebra are two sides of the same coin.

Let's get back to the game itself: the seeds deposited along the squares by the competitors are tangible, manipulable and countable, and are used to separate the players at the end of a game. They symbolise mobility, animation, ascent and regression, and so on. As for geometry, it is embodied in all the shapes in general, the game board and the seeds. However, the configuration of the game in its animation tends to reveal algebra as the vivifying element, and geometry as the object to be vivified. So figures and numbers seem to be accorded a certain superiority over geometric forms. In other words, the life-giving character generally attributed to the triangle pointing towards the heavens is manifested by the seeds passing from one square to another across their common borders. The life-giving nature of algebra over geometry is a relationship that is clearly made explicit in the traditional Songo game.

V-) THE MENORAH AND THE HANUKKIAH

Source :https://www.wikipedia.org>wiki>Hanoukkia

1-) The candlesticks and the Bible

a-) the book of Exodus chap 25:31

All the sacred objects contained in the pages of this book have one thing in common, because they are always linked to the migration story of the people they belong to. So the new nationalities that are wrongly attributed to us are distancing us from our own authentic history, in other words from God. The present situation is mainly devoted to the Jews, or the Hebrews, but later it will be shown that all share the pyramid as a common religious and mathematical symbol. It is worth dwelling on the following passage to understand that the Egyptian side can of course be revealed here:

<< The Menorah was a seven-armed structure on which oil lamps burned, which is described in great detail, even as regards its shape, dimensions and the material from which it was to be made in the Torah, in the book of Exodus. Indeed, when God appeared to Moses, he ordered him, among other things, to create a special object, destined to become the very symbol of the Jewish religion: "You shall make a candlestick of pure gold; the candlestick shall be of beaten gold; its base, its stem, its cups, its buds and its flowers shall be of one piece" (Exodus 25, 31).>>.

The book of Exodus relates the events surrounding the Hebrew people's exit from Egypt by Moses, at the behest of the Lord. It would therefore not be scandalous to see the Menorah as a pyramid-shaped structure, because the chosen one was instructed according to Egyptian knowledge. It is important to understand that Moses could in no way have conceived of the sacred object without putting something of his own into it, i.e. what he had been taught by Pharaoh, the great and all-powerful ruler of Egypt. The Menorah, as a symbol of the Jewish religion, lends even more credence to what has gone before: the Eternal's practical covenant with the Hebrew people through Moses, in a catastrophic situation where this long-suffering people had no idea which saint to turn to. He became an unforgettable memory, a guarantor of freedom, and now that God was with them, who could defeat them? The author's thoughts below are perfectly in line with these words:

<< *According to Zechariah, these seven lamps are the eyes of God watching over all the Earth (this is an interpretation; the 7 eyes seem rather to signify God's omniscience).* >>

Sacred art thus reveals itself as a desire for the invisible to be transposed within a practical and possessive work, by way of authentic testimony. It is more than a mark of identity or a religious symbol, it is a return to the genesis of the Hebrew people. The Menorah is quite simply a marvellous story with no end.

It should not be forgotten that the Hanukkiah or 9-branch candlestick also has its own religious and miraculous history. In spite of everything, it also has a pyramid-shaped framework that needs to be completed as we go along.

b-) the book of Mark chap 8:9-20

The present fall in the New Testament tends to remind the supporters of modernity that nothing needs to be discovered any more, but rather explained in minute detail, because the language of the ancients was very learned. We need to translate the following biblical passage into mathematical language, since it is supposed to be based on the combined frameworks of the 7-branched candlestick and the 9-branched candlestick:

<< When I broke the five loaves for the five thousand men, how many baskets full of pieces did you take? Twelve, they replied. And when I broke the seven loaves for the four thousand men, how many baskets full of pieces did you take? Seven," they replied.

The preceding passage shows the arrangement of numbers in triplet form:

We have:

E1: 5-5000-12

E2: 7-4000-7

Transforming Elen into a 9-branch candlestick and E2 into a 7-branch candlestick.

We have: E1:5-5000-12=5-(5x1000)-5+7, now the Hanukkiah has nine (9) branches, and the triplets 4+(4+1)+4, 444, 454, and 5-5-5, hence 3x5=15. The number 5 would also appear 3 times throughout the candlestick, hence: (3x5)+7= 15+7=22, which corresponds to the number of "letters of the Hebrew alphabet". In the same vein, the number 7 can be decomposed, but the result will always be the same. We have: (3x5)+5+2. The number 5 will now appear 4 times instead of 3, so (4x5)+2=20+2=22. Note in passing that the number 5 symbolises the main branch of the 9-branch candlestick.

Let's consider E1:5-5000-12 and calculate E3, E4 and E5 in relation to the number 5, depending on whether the number 5 symbolises the different geometric figures making up the pyramid (1 square and 4 isosceles triangles), the number 3 defining the triangle as the side face, and the number 4 referring to the base which is a square. We just need to count the number of times the same number appears within each term, and multiply it by its frequency of appearance. However, the number 2 isolated in E5 will not be taken into account in order to return the total of the numerical summary of the 9-point candlestick.

We have: E1:5-5000-12,

We pose:

E3= 3+2-(5000)-(3+2)+7=3+2 -[(3+2)1000)]-(3+2)+(3+2) +2=(4x3)+2=12+2=14

We pose:

E4= 4+1- [(4+1)1000)]-(4+1)+(4+1)+2=(4 x 4)+2=16+2=18

We pose:

E5=5-(5x1000)-5+7=5-(5x1000)-5+(5+2)=(4x5)=20

Let E be the sum of E3, E4, and E5, where E:14+18+20=52 or E:14+18+22=54

107

What's more, we pose:

E1: 5-5000-12 =(5-1)+5000+(12-8)=4+5+4 and 1+8=9

Knowing that the number 5 refers to the main branch of the 9-branch candlestick, and that the number 4 refers to the secondary branches, i.e. 2x4=8, then reducing the number 5 and the number 12 to the number 4 implies subtracting 1 from 5 and 8 from 12, so that adding them together effectively reveals the number 9.

The number 5000 refers to the size of the main branch, shamash.

The number twelve suggests a triangular perception of the number 4: T(n)=n+(n+1)=n which becomes T(n)=n+n+n=3n equals T (4)=4+(4+1)+4=4+5+4 and T(4)=4+4+4=3x4=12. The first mathematical formula proposes a greater perception of the numerical symmetrical centre, while the second formula reduces the symmetrical centre by one unit. The Bible offers a double perception of the centre of arithmetical symmetry, in relation to the ascending triplet of the number 9. The formula is: 4+1+[(4+1)x1000)]+4+8=5+(5x1000)+12=5+5000+12.

Factor in the number in the centre to link the candlestick to the Hebraic alphabet numerically speaking:

E3=3+2-(5000)-(3+2)+7=(3+2)-[(3+2)1000)]-(3+2)+(3+2)
+2=4(3+2+2=(4x5)+2=20+2=22

We pose:

E4=(4+1)- [(4+1)1000)] -(4+1)+(4+1)+2=4(4+1)+2=(4 x 5)+2=20+2=22

We pose:

E5=5-(5x1000)-5+7=5-(5x1000)-5+(5+2)=(4x5)+2=20+2=22

Let E be the sum of E3, E4, and E5, where E=22+22+22=(3x22)=66

E2: 7-4000-7 offers a reading of the branches of the Menorah from right to left and from left to right, i.e. 1,2,3,4,5,6,7 and 7,6,5,4,3,2,1. In this way, the number 7 is found on both the extreme left and the extreme right, while retaining the main branch represented by the number 4, which is multiplied by 1000. This is how the triplet 7+4000+7 is obtained. The triplet is normally relative to the top view of the arithmetic symmetrical centre. In fact, we start from 3+1+3=7 for 3+(3+1)+3=3+4+3=10, which becomes (3+4)+4+(3+4)=7+4+7, the central value 4 is added to the identical values of the symmetrical bounds. T(n)=n+(n+1)+(n+1)+n+(n+1) equals T (3)=3+(3+1)+(3+1)+3+(3+1) = (3+4)+4+(3+4)=7+4+7=18

Adding the numbers in order of higher and lower linear growth and decay gives numerical stagnation at one digit or higher. We pose:

1,2,3,4,5,6,7

7,6,5,4,3,2,1

8,8,8,8,8,8,8

The result simply shows that the number 7 is the interval value of the number 8. In other words, arithmetic triplets whose symmetrical centre is equal to 1 are all interval values of even numbers. For example, if we put 3+1+3=7 and 3+2+3=8, then 7 is inside 8. This is a non-homogeneous parallelism.

The triangular configuration means that the digits 7 form a single vertex, giving an angular and linear correspondence between the digits. The equation is :

1,2,3,4,5,6,7

1,2,3,4,5,6,7.

Triangular duality is the hallmark of the Sacred Harp or Ngombi or Ngoma. In contrast, the Harp Cithara, traditionally known as the Mvet Ekang, the Menorah and the Hanukkiah, share a tripartite perception. One poses:

1,2,3,4,5,6,7

1,2,3,4,5,6,7

1,2,3,4,5,6,7

In reality, there are 8 perforations and not 7, because the bridge is also located on a so-called neutral perforation. This mathematical logic shows that it is the interval value that is made practical at the expense of the practical value. The corresponding triplet is therefore 7+1+7, which becomes 7+(7+1)+7=7+8+7, then one unit is subtracted from the central value, leaving the homogeneous triplet 7+7+7= 3x7=21. If we add cordophones, it becomes 7+7+7+7=4x7=28.

In the case of the 7-branch candlestick, we have: 3+1+3=7 which becomes 3+4+3=10, hence the homogeneous triplet 3+3+3=3x3=9. Also counting from the right towards the centre, from the left towards the centre, and from the centre downwards, we have: 4+4+4=3x4=12

The candlestick has 9 branches: 4+1+4=9 which becomes 4+5+4=13, then 4+4+4=3x4=12, at the end 5+5+5=3x5=15

As in E1, the number 12 can still be seen in E2 through the following: (3+4)+4+(3+4)=7+4+7, or 3x4=12. The number 4 appears 3 times in this numerical triplet. In fact, the idea of the arithmetical triplet revolves around a neutral energy that ultimately unifies beings of the same nature, while standing in their midst. Schematically, Christ is this neutral energy that unites the apostles and the crowd through the word of salvation. The homogeneous triplet 444 or 4+4+4, the sum of which is 12, refers to the light that now houses those who put the gospel of Christ into practice.

The connection between the number 7 and the number 22, indicating the letters of the Hebrew alphabet, is justified by a 7cm long square. In fact, it contains 11 parallel straight-line segments, read from top to bottom and bottom to top, i.e. 11 above against 11 below, giving 11+11=22.

Another trick is possible, by talking about the calculation of addition by order of stagnation and arithmetical assimilation. We pose:

7Stg (4)=1 2 3 4 5 6 7

= 2+(1+3)+4+(5-1)+(6-2)+(7-3)

= 2+4+4+4+4+4

= 2+(4x5)

=2+(5+5+5+5)

=2+20

7Stg(4) = 22

It should be noted that the number 2 is always isolated, as in the calculation of equations E3, E4 and E5, in relation to the number 9, so E1: 5-5000-12. The multiplicity or multitude is expressed in a triangular form in this Gospel, with Christ as the symmetrical centre of the triplet, to which the apostles and the people must obviously be assimilated. Christ is light, and those who follow him also become light. The candlesticks, which bring light and many other sources of abundance to the Hebrew people, are revealed in the words of Christ in the New Testament, while at the same time merging with the letters of the Hebrew alphabet. The number 12 may symbolise Christ's apostles, but its interval value is 11, which explains why he was betrayed by one of them. The number 11 refers to the spirit of truth, the purity of heart, which unfortunately not everyone had. This is why the cordophones of both Ngombi and Mvet refer not only to the inner region of the square, but also to its axes of symmetry and diagonals, to signify that it is the abstract that must obviously take precedence over the concrete work. The spirit must dominate the matter.

c-) the book of Daniel chap 3: 19-25

The section reconciling the migratory narrative and sacred art is still in progress. The return to the heart of the Old Testament relates to the present analytical content. It is important to understand how the 7-branched candlestick or Menorah is revealed in this verse. It is written:

*<< Then Nebuchadnezzar was filled with fury, and the expression on his face changed at the egcird of **Shadrack, Meshak and Abednego**. He spoke up and ordered the furnace to be heated **seven times,** more than was customary. Then he ordered strong men from his army to bind Shadrack, Mëshak and Abed-nego, to throw them into the furnace of blazing fire. Then these men were boundëes with their trousers, their tunics, their caps and their minds, and they were cast into the midst of the furnace of burning fire. On top of this, as the king's word was rigorous and the furnace had been extraordinarily heated, those very men who had hoisted up Shadrack, Meshak and Abed-nego, the flame of the fire killed them. As for these three men, Shadrack, Meshach and Abednego, they fell bound in the middle of the furnace of blazing fire. King Nebuchadnezzar heard them singing; he was stunned and got up hastily. He spoke up and said to his advisers: <<Did we not throw three bound men into the fire?>> They replied and said to the king: <<Of course, d king!>> The king replied and said: <<Behold, I see **four** delirious **men** walking in the midst of the fire with no wounds on them, and the appearance of the **fourth** looks like that of a son of the gods.>> (Daniel 3:19-25).*

First of all, the story is built around 3 individuals, including Shadrack, Meshak and Abed-Nego. Later, we are made to understand that the furnace is heated 7 times more. We know that the candlestick does indeed have 7 branches, and that the 3 branches make up half of them, either on the left side or on the right side. Afterwards, the 3 individuals are thrown into the furnace of fire, where miraculously another, i.e. a fourth, is invited to the trial, which completely surprises King Nebuchadnezzar of Babylon. Evolution offers the passage from 3 to 4 individuals, i.e. the main branch, i.e. 3+1=4. Curiously, this saviour ends up disappearing without explanation, just as he appeared, so we end up with the same 3 individuals we started with, but this time after passing through the fire. So we have 3 individuals before the fire, 4 in the furnace, and 3 individuals afterwards. The triplet 343 or 3+4+3=10, is a magnified perception of the ordinary triplet 3+1+3=7. In view of the above, we can assume that the furnace was heated 10 times, i.e. 3 times more than usual, i.e. 7 times. Remembering the Menorah as a symbol of protection received by Moses from the Lord, we can see that he obviously remained faithful to his promise. We give credence to Zechariah, who said that the 7 lamps represented the 7 eyes of God who watches over the universe. The number 4 therefore refers to the guardian angel, or quite simply to the presence of the Holy Spirit.

2-) candlesticks and the hebraic alphabet

a-) presentation

It is always essential to remember that the observation began with the study based on the framework of the Menorah, and then led to the Hebrew alphabet. In other words, the mathematical analyses carried out on the 7-branched candlestick led to a certain correlation with the Hebrew alphabet. Sculpture is, of course, considered to be a scripture in its own right, capable of revealing unsuspected knowledge. It would therefore not be out of place to offer an intoxicating flavour of this alphabet through what follows:

"The Hëbrew alphabet is much more than an alphabet; it opens up access to universal knowledge by way of the symbol. As human beings, we do not have the possibility of being in direct contact with the invisible. Mystery cannot be communicated in this way, but it is revealed in the form of symbols and metaphors. It is through symbols that we will gradually come to understand the world of mystery. The symbol is a real channel of communication between the visible and the invisible.

The Hebraic alphabet, through the symbolism of each of its constituent letters, has the ability to plunge the reader into a universe where he discovers himself against his will. It not only leads the observer to the very genesis of the world or universe, but also, and above all, reminds him that his own universe is, by its very nature, invisible. It must therefore be constructed within this very invisibility, through its immaterial part. In reality, the human body is like a vehicle driven by a driver, who sometimes has to live outside it. The point here is to create another personality beyond one's own assets, because it can belong to others. The soul has to be saved for eternal life, whereas the body is only destined for a single or limited use. Hence the need to consider the future of life outside the human body.

b-) the 22 letters

The internal structure of the Menorah shows a horizontal alignment of 7 points, symbolising the branches, and an alignment of 4 points on the central plane, all referring to a right angle. So we need to count the 7 branches from my left to the right, then from the right to the left to obtain the number 14. Obviously 7+7=14. Then, as 4 points on the lateral plane from top to bottom and from bottom to top, the number 8 is obtained. Finally, 8+14=22. We should now use the following passage as an illustration:

<<In Hebrew, letters have a numerical value and can be used to count. This is called "guematria", from the Greek "geometria". Alphabetical numbering is based on ten. It uses the 27 letters of the alphabet (22 ordinary letters + the final 5) in the following way:

The first 9 ordinary letters (aleph, bet, guimel, dalet, he, vav, zayin, het and tet) correspond to the numbers 1 to 9;

The next 9 (yod, kaf, lamed, mem, noun, samekh, ayin, pe andtsade) to the 9 tens (10 to 90);

The last 4 (qof, rech, chin and tav) to the first 4 hundreds (100 to 400);

The 5 final letters (kaf, mem, noun, pe and tsade) in the last 5 hundreds (500 to 900).

From the above, it is clear that the Hebrew alphabet contains 22 letters. What's more, this number can be seen in triplet form, i.e. 9+9+4 or 4+9+9 or 9+4+9=18+4=22. By subtracting the number 2 from each number 9, the triplet 9+4+9=22 becomes (9-2)+4+(9-2)=7+4+7=18. Once again, we discover that the 7-branch candlestick is revealed within the very composition of the letters of the Hebrew alphabet. The present situation shows that the Menorah includes the letters from 1 to 400, because they relate to the number 22. It should always be remembered that the triplet 949 symbolising the number 3, relates to the triangle as the life-giving face of the pyramid.

c-) the 27 letters

Still reconsidering the internal structure of the Menorah, we discover that the lateral plane has 4 points, while the horizontal plane has 7 points. The same 7 points are then reproduced exactly on the horizontal plane on the opposite side, so that the side is caught between these two planes, like the letter H.

Next, we fill in the empty space with 4 vertical points, to create a rectangle. Next, parallel lines are drawn to fill in the geometric shape, to emphasise its fractal nature. This produces the following geometric figure:

Luminous panel 7/4

Sources: N'NANG EBANE Sosthene Tresor

We need to count the number of squares in this 7-square lightbox. There are, of course, several ways of doing this:

In the vertical plane, we have:

2+3+2+3+2+3+2+3+2+3+2=(3 x 5)+(2 x 6)=15+12=27

Or

(2+3)+(2+3)+(2+3)+(2+3)+(2+3)+2=5+5+5+5+5+2=(5x5)+2=25+2=27

On the horizontal plane, we lay:

5+6+5+6+5=(2x6)+(3x5)=12+15=27

On the oblique plane, we pose:

2+4+5+5+5+4+2=2+4+15+4+2=6+15+6=(2x6)+15=12+15=27

The different methods of calculation always reveal the number 27 as the common result. This number also refers to the number of letters in the Hebrew alphabet. The following passage deserves a special mention:

<<In Hebrew, letters have a numerical value and can be used to count. This is called "guematria", from the Greek "geometria". Alphabetical numbering is based on ten. It uses the 27 letters of the alphabet (22 ordinary letters + the final 5) of the following aeon:

The first 9 ordinary letters (aleph, bet, guimel, dalet, he, vav, zayin, het and tet) correspond to the numbers 1 to 9;

The next 9 (yod, kaf, lamed, mem, noun, samekh, ayin, pe andtsade) to the 9 tens (10 to 90);

The last 4 (qof, rech, chin and tav) to the first 4 hundreds (100 to 400);

The 5 final letters (kaf, mem, noun, pe and tsade) in the last 5 hundreds (500 to 900).

Let's reconsider the oblique reading:

2+4+5+5+5+4+2=2+(4+5)+5+(5+4)+2=9+9+(2+2)+5=9+9+4+5 = 27. The couple (4+5) marks the ascent towards the summit, while the couple (5+4) marks the fall towards the base, the number 5 designates the summit of the triangle, and finally the number 4 refers to the base. The Hebrew alphabet is a triangular structure, as shown by the Menorah: 9 on the left side of the triangle, 5 on the top of the triangle, 4 on the bottom of the triangle and 9 on the right side of the triangle. Therefore, the unity of vertex and base translates the triplet 999, as the numerical digit of the alphabet, because 3x9=27.

Analysis of the internal structure of the Menorah shows that the number 27 implies the addition of a fourth (4th) element to the original triplet. This is how we evolve from 9+4+9=22 to 9+4+9+5=22+5=27. If you count the numbers as a single unit, 949 equals 111 or 3 and 9945 equals 1111 or 4, so 1111+111=1222=1+(2x3)=1+6=7 equals 3+4=7. This very arrangement of the letters of the Hebraic alphabet is entirely consistent with the unity of the triangle with the square, through the numbers 3 and 4, which are practically represented by the triplet 994, and the quadruple 9945. Taken in this configuration, the Menorah refers to the numbers from 1 to 900.

d-) the Hanukkiah light panel

d-1) the 27 letters

It makes perfect sense to present here the different readings of the light painting of the 9-branch candlestick, which will naturally be more imposing than the 7-branch candlestick. The image below is a clear view of the latter:

Luminous panel 9/5

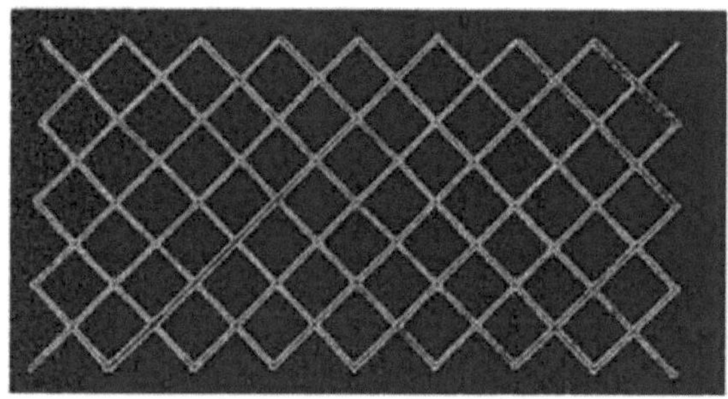

Sources: N'NANG EBANE Sosthene Tresor

We need to count the number of squares contained in this luminous array of 9. There are, of course, several ways of counting:

In the vertical plane, we have:

3+4+3+4+3+4+3+4+3+4+3+4+3+4+3=(3x8)+(4x7)=24+28=52

Or

(3+4)+(3+4)+(3+4)+(4+3)+(3+5)+(3+4)+(3+4)+3=7+7+7+7+7+7+7+3=(7x7)+3=49+3=52

On the horizontal plane, we lay:

7+8+7+8+7+8+7=(8x3)+(7x4)=24+28=52

On the oblique plane, we pose:

2+4+6+7+7+7+7+6+4+2=2+4+6+(4x7)+6+4+2=6+6+28+6+6=(2x6)+28+(2x6)=12+28+12=52

Having obtained the above results, it is now possible to adapt them to the number 27, referring to the number of letters making up the Hebrew alphabet. Using oblique reading, consider the numbers 2+4+7+6+2, then add each of the numbers 2

114

with each of the numbers 7, subtracting one unit (1) from the number 6. (2+7)+(7+2)+4+(61)=9+9+4+5=9+9+9=3 x 9=27.

d-2) The 28 letters

Like the Menorah, the Hanukkiah has a horizontal line made up of 9 points, while the vertical line is organised around 5 points, all referring to a right angle. By counting the horizontal order from right to left and left to right, we get the number 18. Just as the central axis starts with five points from top to bottom and another five points from bottom to top, hence the number 10. So, 10+18=28 or (2*5)+(2x9)=10+18=28. We can see that 6 units have been added to the number 22 in number 7, i.e. (2x4)+(2x7)=8+14=22, to arrive at the number 28 in number 9. In reality, the number 6 corresponds to the triplet 123, i.e. 1+2+3=3+3=6, or half the number 12. Let's look at the following inverse triangles:

Fig 3: Mummies and pyramid

Sources: https//:www.mummies2pyramid.com

The triplet 123, whose sum refers to the number 6, is inside the triangle. However, the triplet 345 outside the trilateral figure is equal to its double, i.e. 3+4+5=7+5=12. It is important to remember that the number 6 designates the interval value of the number 7, which then reveals the number 14 as the external numerical value. The evolution of even numbers is also accompanied by the numerical evolution of odd numbers, but the largest number remains so by just one small unit. The number 6 symbolises the addition of a right triangle with the triangular half of the square, while the number 7 refers to the unity of a right triangle with a complete square as its base. However, rather than the purpose of the mathematical organisation centred around the pyramid being built around the vision relating to the expression of the invisible or the abstract, the number 6 is put forward in relation to the number 7. Just as the number 28 is one greater than the number 27, so the number 7 is one greater than the number 6. The luminous panel of the Hanukkiah offers an excedent perception of this triangle, i.e. 7+8+9=24, which is double the number 12. The number 6 is thus multiplied by 4.

To a certain extent, the two tables also relate to a fourfold perception of the number 12. Consider the first two digits of each set, then add them together.

115

2+3+3+4=2+(2*3)+4=2+6+4=12. There is no doubt that Hebraic candlesticks are full of pyramidal structures.

The number 28 represents both the end of the reading of the Hebrew alphabet and its new reading. The mathematical symbol n+1 implies a return to square one, as n=27, and marks the end. On the contrary, renewal is expressed by the number 28, i.e. the passage from Tav to Aleph. By the same token, the 7th day is Sunday, while the 8th is Monday. The former expresses saturation, while the latter brings forward reinitialisation.

d-3) the letters Yod and Tav

Observation of the joint morphologies of the 7- and 9-branch candlesticks, through the arrangement of the cups in relation to the stem, seems to reveal two main alphabetical letters, Yod and Tav. With this in mind, we should first consider only the cuts on the extreme left and right, and the stem, to highlight the letter Y. It is in the same order that the following passage rightly appears:

<< A Y variant of this shape was drawn by Maimonides in one of his midrashic Iraik's: three straight branches to the east, three to the west and one in the centre. This Y-shape, according to the Kabbalah, was meant to recall the Seven Branches of the Nile Delta; and its holy oil, the sacred waters of the Nile that should never fail. The Y-shape was also adopted by the Lubavitch movement for its Hanukkiah.

The previous passage links sacred art more closely to the Egyptian region, but we must remain within the strict alphabetical framework. The modern letter Y seems to refer to the letter Yod, the tenth letter of the alphabet, whose symbol is the hand and the seed. In the light of the above, the primary meaning of this symbolism is agricultural, as candlesticks are sometimes depicted with the main branch higher than the other branches.

others, and sometimes they are all located at the same level. From these two developments, we can deduce that when the branch is high, it is the harvest season. However, once it is aligned with the others, it defines the sowing season. Fertility, fecundity, is a reality expressed by the candlesticks through their Y variants. In the same way as the fluctuations of the Nile, we can see in it both the period of high water and the period of low water. All the evidence suggests that the candlesticks are abundantly moistened in order to believe in an exponential trend.

What's more, the letter Tav, which is the last letter in the Hebrew alphabet, can be transposed here. Consider the flat trajectory of the branches at the top and just the stem below. So we need to bear in mind the right angle formed by its internal structure. So let's see what the symbolism of this letter refers to:

"The last letter of the Hebrew alphabet, it represents the culmination of Creation and the totality of all things eroded. It is the culmination of a whole teaching, a whole initiation, a step by step towards perfection. Tav is the summary of everything in everything, the integral science of the absolute, the mystery that comes directly to the soul. It is also the cross symbolising the whole and the end of the path. Tav is the absolute, the perfection of creation, allowing the dynamic

breath of the 'Shin' to produce the diversity of forms. It is truth, perfection and the end of the path. This major letter maps out any possibility of postponing an act, and makes the future clear. As a symbol, Tav represents the vibration through which the unspoken and the unspoken are liberated, it is the integration of original consciousness, the end, the fulfilment, the culmination".

In the light of the above, vibration is attributed to the letter Tav as the very expression of life, even though it is the last letter of the Hebraic alphabet. This seems to reflect the idea that true life is imperatively future, where hope is an attitude that ultimately imposes itself as a golden rule. It is the very summary of the sum of each of the letters, which suggest self-knowledge with a view to a new birth. When talking about the pyramid, we have sometimes defined it as the expression of multiplicity within unity, or of unity within multiplicity, because it can be both coupled and decoupled. In the same way, humanity must accept itself as a bipartite subject, in which the invisible must be more dominant over matter. So the Tav itself is full of the idea of overcoming material understanding to devote ourselves wholeheartedly to spiritual work. The destruction of matter for the benefit of the freedom of the soul under the influence of God.

The impulse of the divine is the message conveyed by this alphabetical letter. So its summation with the Yod, based on the ability to regenerate through the hand and the seed, ultimately relates to the idea that the divine's own cannot be extinguished. The oils are rightly used to keep the flame of the cups alive for ever. Just as the Shamash is lit first to signify that life is eternal in a specific place, and is also distributed to us from there. Humanity embodies a secondary life or is simply a recipient of it, whereas the Eternal One is life. We can even see an electrical transposition between the devices (humanity) that receive the electric current, and the generator that produces it (God).

3-) combined light panels

a-) Nature of digits 7 and 9

The numbers 7 and 9 are part of what we call unitary arithmetic triplets, because they have the same centre of symmetry as the number 1 in relation to its identical symmetric limits. It is useful to remember that the first is two (2) units smaller than the second, or that the second is also two (2) units larger. In this way, the same numerical value, which is 2, rises and falls. So, addition and subtraction are presented in a triangular perception through the numerical value marking the difference between two or more successive digits and numbers. The number 7 comprises a regressive triplet rated 3+1+3=7 and an ascending triplet rated 3+4+3=10, while the number 9 relates to the triplets 4+1+4=9 and 4+5+4=13. In view of the above, the triplets of the number 9, by adding their different results, relate to the number 22=9+13. In reality, the ascending triplet of the number 9 reveals the number 13 as its result, but it has the number 7 as its centre of symmetry. In fact, 6+1+6=13, and the central number symbolises the central branch, including the seventh (7). What's more, the number 7 is ranked third (3), while the number 9 is ranked fourth (4), so by simply adding their ranks together, the number 7 still stands out. What can we say about the degree of stagnation 7777 of the light panel of the Hanukkiah? What about its horizontal reading 7878787? The numbers 7 and 9 are, of course, both odd, but the former remains predominant over the latter. The number 7878787 is made up of the triplet 888 and the quadruple 7777, symbolising the numbers 3 and 4, and at the same time referring to the postures of the numbers 7 and 9. So the number 7878787 can also be read as 9797979, by addition: (3x8)+(4x7)=24+28=52 and (3x7)+(4x9)=21+36=57. In view of the above, the number 52 has increased by 5 units, in accordance with the arithmetical centre of symmetry of the number 9.

b-) the 7/4 and 9/5 luminous panel

The complementarity of the Menorah on the one hand and the Hanukkiah on the other, continues until a unique luminous picture is created. The 9-branch candelabra (9/5) is the length, while the 7-branch candelabra (7/4) is the width of the rectangle. The luminous painting taken as a triplet construction refers to a rectangle. The following is a perfect illustration of the above:

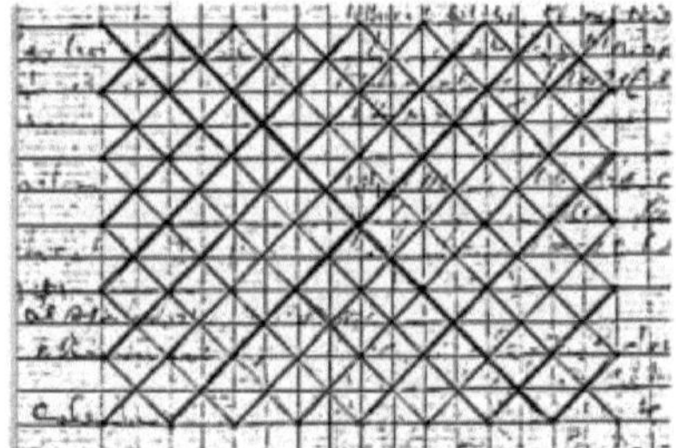

Sources: N'NANG EBANE Sosthene Tresor

We need to count the number of squares contained in this combined luminous panel. There are, of course, several ways of doing this:

In the vertical plane, we have:

5+6+5+6+5+6+5+6+5+6+5+6+5+6+5=(8x5)+(7x6)=40+42=82

Or

(5+6)+(5+6)+(5+6)+(5+6)+(5+6)+(5+6)+(5+6)+5=11+11+11+11+11+11+11+5=(7 x11)+5=777+5=82

On the horizontal plane, we lay:

7+8+7+8+7+8+7+8+7+8+7=(7+8) x 5+7=(15 x 5)+7=75+7=82

On the oblique plane, we pose:

2+4+6+8+10+11+11+10+8+6+4+2=2+28+22+28+2=30+22+30=(2x30)+22=60+2 2=82

In the light of the above, it should be noted in passing that the candlesticks stagnate on the double number 11, the sum of which is 22, which are respectively the interval values of the numbers 12 and 23. In fact, it is possible to extract the number 23 by adding together the vertical dualities of the three luminous arrays.

118

We have 7(2;3), 9(3;4), and 7,9(5;6), from which we get: (2+3)+(3+4)+(5+6) = 5+7+11 = 23. The number 12 can be seen through the triplets of the candlesticks, 3+1+3 which becomes 1234 from the right to the centre, 1234 from the left to the centre, 1234 from the centre to the bottom, hence the number 444=3x4=12. And since the triplet 4+1+4 can be read as 4 from the right, 4 in the centre and 4 to the left, then 444=3x4=12, or 12+12=24. There's another trick to finding the number 24 through the unity of candlesticks. We know that the number 7, despite being an odd number and a unitary arithmetic triplet, is also an interval value for the number 8, which in turn is an interval value for the number 9. So the triplet 789 is obtained by addition: 7+8+9=(7+8)+9=15+9=24. It is worth pointing out that the luminous array of 9 is full of the triplet 789. The numbers 7 and 8 are both considered to be interval values of the number 9. Even in this arrangement, the notion of triplicity necessarily implies the central location of a neutral element, between two elements of the same nature. The digits 7 and 9 are odd numbers, but the digit 8 is an odd number, and is symmetrically linked to them by a descending value of +1.

c-) the 5 internal geometric figures

It is important to note that the number 5 is a common feature of the Menorah and the Hanukkiah. It represents the stagnation point of the 7/4 light table, while at the same time designating the main branch of the Hanukkiah, whose ascending triplet is 4+5+4=13. In terms of the ranking of arithmetical triplets, it occupies second (2nd) place after the number 3. However, what does it symbolise within the luminous array?

Luminous panel 17

Source: NNANG EBANE Sosthene Tresor

It is possible to see from this luminous picture that the most prominent geometric figures are in fact 5. There are four triangles (4), each with a square (1) at its centre. So the number 5 is a numerical perception of the geometric figures that make up the pyramid. It's important to note, however, that the square is a neutral geometric figure in relation to the triangles that surround it, like the numerical triplets. In this case, the number 17 has pride of place, i.e. 8+1+8=17. The number 1 therefore symbolises the square, while the numbers 8 designate the trilateral figures. The luminous painting, through the number 5, offers the vision of an open pyramid.

4-) Special features of illuminated panels

a-) odd numbers

The configuration of the lightbox, based on an odd number, corresponds exactly to a triangle with one vertex. In fact, several vertices make up the luminous array, but only one is located at the centre, in accordance with the structure of the arithmetical triplet used as a construction scale. It should be noted that the trilateral figure occupies this central position at both the upper and lower levels. The following illustration is built around the number 7, whose corresponding triple is 3+4+3=10, or the ascending triplet:

The illuminated 7

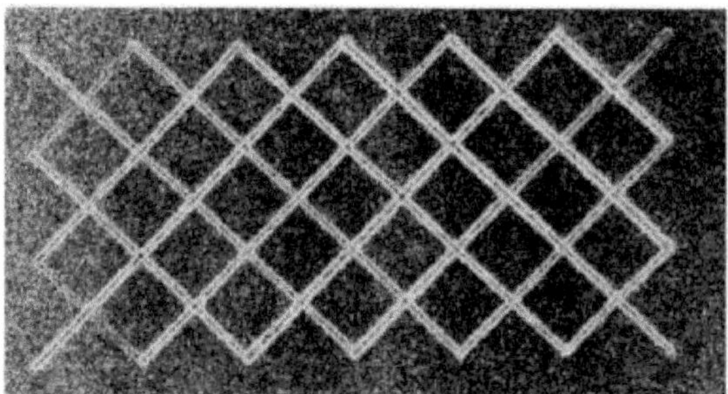

Sources: N'NANG EBANE Sosthene Tresor

In a light painting, the triangle whose sides pass through the right angles is called the principal triangle and is always at the centre of the quadrilateral. Its sides seem to merge with the diagonals, passing through the right angles, hence the term "diagonometric triangle".

The rectangle newly obtained as a container comprises two sides with lengths of 7 perforations and two sides with widths of 4 perforations. It is clear that the rectangle whose interior is completely empty relates to an exclusively personified analysis. In other words, a study outside the group that is the pyramid. However, the representation of the diagonals of the quadrilateral connects it to the pyramid in a simplified perception, resulting in a unification of the triangular vertices at its centre. The vertices of the triangles in the luminous array point to the centre of each longest side, so by simply tracing them vetically, we discover them as "axes of symmetry", which at the same time constitute the diagonals of the vertical square or of the square that is surrounded by triangles. There is therefore a connection between the triangular vertices, the axes of symmetry, and the diagonals, within the rectangle, which is the product of the arithmetical triplet, and therefore of both central and orthogonal symmetry. The arithmetical triplet is an arithmetical perception of geometrical properties, such as parallelism, central symmetry, perpendicularity, fractal geometry, etc,. In short, the triangle that points towards the heavens will now point downwards, and the one with the lower vertex will now point upwards, so that the two vertices are joined at the centre of the rectangle or at the centre of both the diagonals and the axes of symmetry.

120

Triangles, like rectangles, have 7 holes on each side, which makes them equal, and each has 15 squares. These are the triangles that extend downwards and upwards, not those that surround the square.

"If in a given geometrical figure the length of the cords of the opposës triangles is equal to the length of the cords of the latter, then it is a rectangle".

The square appearing at the centre of the equilateral triangles has 4 perforations of length on each side, i.e. 4444, and has 9 squares. The 7-perforation square for the number 7777 is reduced to a 4-perforation square for the number 4444, a reduction of three units (3). Starting from the centre of the square, i.e. the diagonal for the vertex, the base has 4 perforations, the sides of the rectangle each have 4 perforations, giving the triplet 444, the sum of which is 12. So the upper 444 triplet is opposed to the lower 444 triplet, but each number 4 in the centre relates to the diagonal.

If a square is located at the centre of the union of two inverted ëquilateral triangles whose vertices point to their respective opposite bases, then we speak of "trilateral framing on a quadrilateral". The lengths of the sides of the 'quilateral' triangles are greater than the length of the square of 3 respective units. This is how the "regenerative diagonometry" is revealed.

The horizontal right angles of the tile are linked to the vertices of the resulting horizontal equilateral triangles, each made up of 3 tiles. It has to be said that we are witnessing a framing of the order of three through this mechanism, i.e. 3<9>3, from which we get: Tn=n+(n*n)+n equals 3+(3*3)+3=3+9+3=15.

If the number 7 is used as the basis for constructing a rectangle via its tripet, then the triangles within it have a single vertex and point to the numerical value 15, as the degree of stagnation. The corresponding triangular numerical condensation is: 2+4+5+5+5+4+2=27. The number 5 appears three (3) times, marking the stagnation of the triangle before later regressing from 4 to 2, i.e. from vertex to base. If we look at the triad of the sun in the sky, for example, we can obviously assume that the numbers 2 and 4 designate its rising, the number 15 its zenith, and finally the numbers 4 and 2 refer to its setting. This means that the solar star heats up the most at the zenith, which is its highest degree.

By analysing the diagonals, we can understand the different interdependent relationships between the square on the one hand and the triangle on the other, which behave like Siamese twins in condensed and fragmented forms. The trilateral figures are closed to the quadrilateral on one side, before later opening up to the quadrilateral on the other (a situation of triangular rotation around the square).

b-) even numbers

The even number chosen for this illustration is 8, whose corresponding triplet is not very different from that of the number 7. The latter has one unit in the centre, while the former has two units in the middle, giving 3+1+3=7, compared with 3+2+3=8. For even numbers and digits, T(n)=n+2+n is equivalent to T (3)=3+2+3=8 or T(4)=4+2+4=10. The corresponding ascending triplets are 3+(3+2)+3=3+5+3=11 and 4+(4+2)=4=4+6+4=14. "The direct impact of even numbers is that the sides of the triangle passing through the right angles with the vertex are not unified, and tend to be lost between parallel or individual vertices". A pictorial perception of the above is not at all misleading:

Luminous panel 8

Sources: N'NANG EBANE Sosthene Tresor

If we look very carefully at the trilateral figure, we can see that the sides passing through the right angles do not have uniform vertices. However, we can highlight the presence of a triangle with a vertex in the middle at the bottom level, but its sides do not pass through the right angles of the quadrilateral.

The triangle with a single vertex has 10 squares rather than 15, and the emphasis is now on an even number. If we consider the three triangles, we can see that the number 5 is arranged fourfold. For example, $T(n)= n+n+n+n$ or $n+(n*2)+n$ equals $T (5)= 5+5+5+5=4*5=20$ or $5+(2*5)+5=5+10+5=20$. The number 7 offers a threefold perception of the number 5 through the number 15, while the number 8 reveals a fourfold arrangement of the number 5, the sum of which is 20. The number 7 evokes the number 3, while the number 8 relates to the number 4. However, the more the former is inside the latter, the more the number 8 will evoke both the unity of the number 3 as a symbol of the number 7, and the number 4 as its own regressive representation. This is why the Sacred Hapre has 8 upper perforations, 8 cordophones and 8 lower perforations. Secondly, it also has 8 vices, 8 superior perforations, 8 cordophones, and 8 inferior perforations. So the triplet 888 represents the number 7, while the quadruple 8888 represents the number 8. It was later discovered that the numbers 7 and 8 are internal arithmetic values of the number 9, from which the Ngoma, Ngombi or Sacred Harp is derived. In this same perspective, the number 7 and the number 9 occupy the same rank in terms of odd numbers that can be constructed in triplet form, i.e. the 3rd and the 4th. The number 4 is thus common to the numbers

8 is its half 4+4=8, and the number 9 is its two symmetrical terminals 4+1+4.

In this configuration, the diagonals form four (4) equilateral triangles with two different vertices: two (2) above and two (2) below, two (2) to two (2). This is what is known as 'diagonometric triangular collapse'. This mathematical phenomenon is also known as "Triangular Shading in M or W". In effect, there are two parallel vertices located above one another in the middle in a lower region.

It is observed that only even numbers produce the triangular triapatite effect or diagonometric triangular shading, because they have a centre of equilibrium different from one (1).

If a triangle has two parallel vertices and an inferior vertex, then this is a diagonometric triangular shading situation, which is only made possible by using half a digit or even number as the length of the square.

The tripartite triangle has a total of 20 squares, just one less than the trilateral figure on the quadrilateral. However, there is still a tripartite frame expressed in it. T(n)=n+(nx2)+n is equivalent to T(5)=5+(2x5)+5=5+10+5=20. Alternatively, T(n)=4*n is equivalent to T(5)=4x5=20, corresponding to the degree of stagnation of the triangle. The corresponding numerical triangular condensation is: 2+4+5+5+5+5+4+2=6+20+6=32. The numerical symbol of the vision remaining on the vertex of the triangle is: 6<20>6.

In the light of the above, the number 8 merges with the number 7 because, as it is not an odd number, the same square with 4 perforations of length appears twice in the reduced situation. T(n)=n+[(nxn)+(nxn)]+n equals T(3)= 3+[(3x3)+(3x3)]+3=3+9+9+3= 3+(9+9)+3=3+18+3=24, by extension: T(n)=n+(n+2)+(n+7)+(n+2)+n equals T(3)=3+(2+3)+(3+7)+(2+3)+3 = 3+5+10+5+3=(3+5)+10+(5+3)=8+10+8=26. In reality, to get the exact number of squares, we need to add the number 3 to each end of T(5). T(n)=n+[(3x(2+n)]+n equals T (3)=3+[(3x(2+n)]+3=3+[(3x(2+3)]+3 = 3+(3x5)+3=3+15+3=21.

Fig: Diagonometric triangular shading

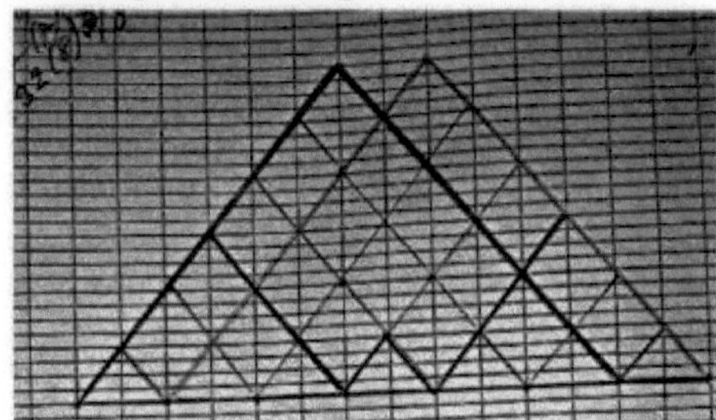

Sources: NNANG EBANE Sosthene Tresor

In this configuration, the square at the centre of the trilateral figures is counted twice (2). Hence, T(n)=n+(nx2)+n is equivalent to T(n)=n+(nxn)+(nxn)+n= T(3)=3+(3x3)+(3x3)+3=3+9+9+3=3+(9+9)+3=3+18+3=21. A triangle contained in a square whose length is 8 perforations contains 21 squares. Note that diagonometric triangular shading actually unifies two triangles that are equilateral, but become isolated with respect to the length of their common base.

If a triangle has two parallel vertices and one inferior vertex, and the length of their common base is greater than the length of one of their sides, which are equal two by two, then it is a diagonometric triagular shading.

If a triangle has two parallel vertices and a central lower vertex, then the square within it is also naturally double, with exactly the same dimensions.

Diagonometry allows us to understand the possible mutation of an isosceles triangle into an equilateral triangle, and vice versa, using the diagonal as the axis of symmetry, by means of parallelism and the fractal arrangement of the line segments parallel to it.

Odd numbers refer to triangles with a single vertex in height, while even numbers are synonymous with triangles having two parallel vertices and an internet and lower vertex.

The triangular diagonometric shading simply expresses the passage from unity to duality. A single element capable of regenerating on itself, with the obvious consequence of giving birth to its own double. Diagonometry shows beyond any doubt that neither the square nor the triangle can really be independent of each other.

The triangle can be carried by the square, just as the trilateral figure carries the quadrilateral in a vertical arrangement. We'll come back to this in a later analysis. This observation contains a very profound idea about time, according to which everything created is necessarily destined to come to an end, because life is a cycle. Today's dominants will be tomorrow's dominants, and vice versa.

The table below illustrates the numerical values of the triangle, depending on whether the digits and numbers are even or odd:

The evolution of cords within the frame

Geometric figures	Carre					Triangle
Numbers	Internal Value 1(VI1) n-2	Internal value 2 (VI2) n-1	VI1+VI2	VI1XVI2	Multiplication by 2	Divion par 4
1	-	-			-	-
2	0	1	1	1	2	0,5
3	1	2	3	2	4	1
4	2	3	5	6	12	3
5	3	4	7	12	24	6
6	4	5	9	20	40	10
7	5	6	11	30	60	15
8	6	7	13	42	84	21
9	7	8	15	56	112	28
10	8	9	17	72	142	35,5
11	9	10	19	90	180	45
12	10	11	21	110	220	55
13	11	12	23	132	264	66
14	12	13	25	156	312	78

15	13	14	27	182	364	91
16	14	15	29	210	420	105
17	15	16	31	240	480	120
18	16	17	33	272	544	136
19	17	18	35	306	612	153
20	18	19	37	342	684	172
21	19	20	39	380	760	190
22	20	21	41	420	840	210
23	21	22	43	462	924	231
24	22	23	45	506	1012	253
25	23	24	47	552	1104	276
26	24	25	49	600	1200	300
27	25	26	51	650	1300	325
28	26	27	53	702	1404	351
29	27	28	55	756	1512	378

Sources: N'NANG EBANE Sosthene Tresor

The table shows a sequence of arithmetic triplets, taking into account the column of data VI1+VI2. It can be seen that their different results relate to the odd numbers taken as arithmetic triplets. The data goes from 1 to 55, and evolves according to the order of growth of two (2) units. However, it is advisable to consider only the data in the column for division by 4, in order to discover which number of squares a given digit or number corresponds to, depending on whether it is even or odd. Example: the number 7 is linked to the number 15 (number of squares and degree of stagnation), the number 8 to 21, the number 9 to 28 (number of squares and degree of stagnation), and so on.

5-) the hexagram

It is important to continue with the rectangle obtained above, based on the internal framework of the Mvet Oyeng instrument, which is also common to Jewish candlesticks. Religious instruments all have a right angle, giving them an alphabetical shape referring to the letter T. The Mvet's shape describes a right angle. The shape of the Mvet depicts an upside-down T, while the shape of the candlesticks reveals its ordinary posture. The triangular sketch that follows is taken from a 9-branched candlestick:

Figure: a rectangular sketch

Source: NNANG EBANE Sosthene Tresor

You will need to draw two inverted triangles, the first of which will point towards the heavens, while the second will point downwards. The vertices of the trilateral figures are fixed on the 'centres' or 'middles' of each of the lengths of the quadrilaterals. Remembering the numerical triplet 4-1-4, the sum of which is 9, and 4-5-4=13, the numbers 1 and 5 designate the centre of each length according to the mathematical formula used: i.e. $T(n)=n+1+n$ equals $T(4)=4+1+4=9$ and $T(4)=4+(4+1)+4 = 4+5+4=13$. The following result leaves us speechless:

Fig: 1'hexagram

Source: NNANG EBANE Sosthene Tresor

You can see that there are two inverted triangles inside the rectangle. The length of the rectangle above is 17 cm and the corresponding triplet is 8-1-8, while the width measures 9 cm and the triplet is 4-1-4.

The resulting inverted triangles are not dissimilar to the Star of David, known as one of the emblematic symbols of the Jewish religion:

"From a purely aesthetic point of view, it corresponds to what we call a hexagram. This gëomëtrical form could be written by the superimposition of two triangles, one pointing upwards, the other pointing downwards".

The Mvet instrument, with its triangle pointing towards the heavens, is therefore half of the hexagram, as the triangle pointing downwards is not represented at all. In the opposite direction, the triangles of the candlesticks only point downwards, because the cuts, by their alignment, symbolise the base of the triangle, constituting at the same time one of the diagonals of the square, hence the term "diagometric triangle". It is therefore possible to find geometric figures that resemble the latter,

126

undoubtedly to amplify once again the mysterious aspect of sacred art. It is also known as the 'Shield of David', just as the Ekang Mvet is a protective shield for the Ekang people.

There are 12 triangles in all, 10 of which are formed by the main inverted triangles, while forming a square in their middle, which can be divided into 4 triangles. There are thus 11 geometrical figures present within the hexagram, 10 triangles and 1 square. There are therefore 10 triangular correspondences in triplets, from the smallest value to the largest. We have: 111, 222, 333, 444, 555, 666, 777, 888, 999, and 101010.

The triangle of candlesticks is double that of the Ngombi, because it is divided into two parallel planes of correspondence, like that of the Mvet Oyeng. On the left, the 4 horizontal cups are linked to the main branch by 4 junctions, the whole linked by 4 branches. On the right-hand side, there are 4 cups, 4 junctions and, of course, 4 branches, so 444=(3x4)=12 and 444=(3x4)=12, and 12+12=(2*12)=24. For the candlestick with 7 branches, 333=(3x3)=9 and 333=(3x3)=9, and 9+9=(2x9)=18.

Excluding the square contained within the triangular formation by contrast, it is possible to establish a mathematical relationship with the image below. Another arithmetical relationship can be established through the hexagram obtained, to better explain the image of the mummy standing close to a triangle having the triplet 123 inscribed within it and the triplet 345 lying outside. An image view is of the utmost importance:

Fig 3: Mummies and pyramid

Sources: https//:www.mummies2pyramid.com

Each triangle is the product of three (3) aligned points, so the sum is 6, just as 1+2+3=6. What's more, uniting the four (4) triangles with the square (1) at the centre reveals a geometric figure with 12 sides. Each triangle has two (2) sides (2x4=8), but their bases coincide with the four (4) sides of the square (4+8=12). In other words, the triplet 231 symbolises the triangle that points towards the heavens, while the triplet 534 symbolises the triangle that points downwards. The hexagram is thus built around the number 12, which is a reading centred on one of the parts of the 9-branch candlestick.

The hexagram is the product of fractal geometry constructed by the parallel and perpendicular correspondence of an arithmetical triplet taken as the scale of construction within a virtual rectangle, the whole of which relates to the image of an open pyramid, more precisely the unity of two main triangles whose vertices point to the centre of each of the opposite bases, giving rise to four triangles and a combined square.

The hexagram offers a metamorphosis of the diagonals of the square into the lengths of the sides of the triangles, whose vertices play the role of axes of symmetry. The diagonals of the square are thus reconverted into axes of symmetry, and then the axes of symmetry into diagonals of the square. This is only perceptible by completing the geometrical figure, as the candlesticks and the Mvet Ekang propose either a superior or inferior face of the right angle constituting their common skeleton. The hexagram is also a geometrical figure seen in half.

6-) The vertical edge

The analysis stage always leads to an interest in the different geometric shapes inside the rectangle. After the inverted triangles, it's the square's turn to be discovered. This is no longer an ordinary square, but a so-called vertical square. It would certainly be easy to understand that it is its position between two inverted triangles that favours this configuration.

It was said above that the ordinary tile is covered by four triangles. The value of the length of the side noted $V(n)$, is superior by two (2) units to the first value of the growth of the tiles within it noted $VI1(n)$ and by one time superior to the second value of the order of evolution of the internal tiles noted $VI2(n)$. The quadrilateral, by contrast with the arrangement of the cordophones, allows the tiles to evolve on a descending and ascending curve to infinity. The candlesticks also conform to this configuration.

On the contrary, the vertical square knows no internal passage from the smallest of values to the largest, but rather an infinite constancy. An examination of the geometric figure is highly commendable:

Fig: 1'hexagram

Source: NNANG EBANE Sosthene Tresor

128

Let's go back to the construction scale. The length of the rectangle is 17 cm, and the regression triplet T (n)=n+1+n equals T (8)=8+1+8=17 and the ascending triplet T (n)=n+(n+1)+n equals T (8)=8+(8+1)+8=8+9+8=25. The vertical edge is symbolised in both cases by the numbers 1 and 9, but to find out the length of its sides, we need to consider the midpoint of T(8)2, i.e. the number 9. So V(n) is 9 cm but VI(n) is 8 cm, demonstrating once again that the second number is expressed within the first, in memory of the Ngombi triangle. The length of the tile is therefore 9cm, but the number of tiles between them is 8 per square. VI(n)=(Vn-1)$^\pi$ 2 equals VI(n)=(9-1) 2=8^{AA} 2=8*8=64. The number 64 denotes the sum of the tiles in a vertical tile whose sides are 9 cm long.

It is therefore the numerical triplet that favours the position of the vertical square within a hexagram, which is inscribed within the rectangle. It is advisable to recognize that all is built around the figure 3, the square is internal with the triangles, which in their turn are inscribed within the rectangle. The hexagram necessarily implies a tripartite composition of particular geometric figures, hence the expression of multiplicity within unity.

In addition to the rectangle, the triplet can also be used to create the hexagram. The square is a direct consequence of the six-pointed star. The ordinary square has two internal values (-2) and (-1), while the vertical square contains a single internal value (-1). The former refers to 'decay'(\) and 'growth'(/), while the latter denotes 'stagnation'( ).

A single numerical value is continuous outside the tile, including from the inside this time. Stagnation is in full swing in this layout, with inner thought expressed from the outside. The vertical square thus represents the domination of matter by the mind, the stage of maturity, the ability to channel energies in order to achieve a precise goal. This posture of the geometrical figure fits in perfectly with the term Mvet designating elevation, and the luminosity highlighted by the candlesticks, which in reality involves enormous sacrifices in order to bring to fruition a project that we hold dear to our hearts.

The form is not just to be traced without understanding the idea behind it. The eyes of some are limited to the instrument as a simple folk object, which is an obvious mistake to be pitied. The posture of the square, barely standing on the ground, is similar to a man walking on tiptoe to maintain his height, and the name of the sacred art also reveals this majestic aspiration. The detachment of the soul from material possessions is clearly expressed here, and the need for humanity to consult the world of implacable intelligences is a melodious message brought to the fore by this symbol.

7-) The square triangles

The hexagram inscribed within the rectangle now makes it possible to understand the true nature of the triangles in the traditional Menorah, Hanukkiah and Mvet Ekang. It is no longer based on the type of triangle, but on its components, with reference to the hexagram. Let's take a closer look at the geometric figure:

Source: NNANG EBANE Sosthene Tresor

The triangle on the candlesticks points downwards, whereas the triangle on the Mvet points upwards, which is why the two together form the hexagram. Naturally, it is important to consider only the upward-pointing triangle, which means that the instrument is represented as half, just like the Ngombi triangle, which is actually half of a square. It is therefore ascent that is revealed as the primordial idea.

Let's go back to the composition of the trilateral figure, which comprises a "vertical square" located above "two isolated triangles", i.e. one to its right and the other to its left. There are thus three (3) triangles, the largest of which has two others of equal size inside it. From this point of view, the emphasis is on elements of the same nature, which refers to the half calabashes (3) of the same portion, and the cups of the same size. On the other hand, the posture of the vertical square between two isolated triangles refers to the unity of opposite things, which corresponds to the large half calabashes between two others of the same size, just as the Shamash is higher than the auxiliary branches of the 9-branch candlestick. Squares and triangles are all geometrical figures, but they differ in nature. The quadrilateral is arranged like a bird with its wings outstretched and its legs unified. One of its four right angles touches the base of the large triangle, while the three remaining angles remain high. The image of the Virgin Mary and that of Jesus Christ are effectively contained within the virtual vertical squares. The following is a perfect illustration of the above:

Fig: a triangle bursts open

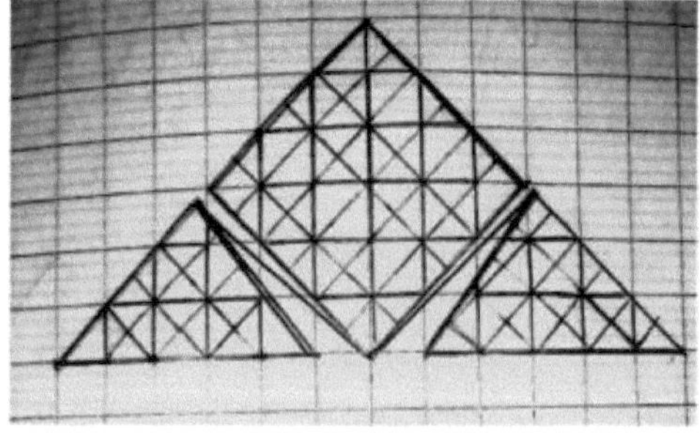

Source: NNANG EBANE Sosthene Tresor

The vertical edge is located exactly above two isolated triangles of the same size. The quadrilateral therefore appears more imposing than the trilateral figures taken together. The number three (3) is recurrent in sacred art, not only to indicate the three sides of the triangle, but also to designate the geometric figures that make it up. The number thus evokes both homogeneity (external triangle (1) and internal triangles (2)) and disparity (square (1) and triangles (2)), a single element adopting several internal configurations. A triangular multitude, compact from a still view, and dispersed (mobile) through its internal dynamism. Everything leads us to believe that it is the internal configuration of the overall triangle that is thus brought to the fore by the study. From the foregoing, it is clear that the square is a "quadruple triangle", associated with the internal triangles (2), so the triangle contained within the candlesticks is revealed in 7 dimensions. The overall triangle (1), the triangles making up the square (4), and the trilateral figures below the quadrilateral (2). The triangle is thus revealed as the dominant geometric figure, because it makes up the quadrilateral. So the 7 branches of the candlestick or Menorah can also refer not only to the triangle as the most prominent geometric figure, but also to its internal components, including the square and two other triangles. The configuration of the geometrical figure and the flame contained within the cups bring to the fore the idea that it is the content and not the container that should be valued. This content is no more and no less than the expression of the invisible, or quite simply a symbolic expression of the divine. Civilisational memories describe the metamorphosis of the triangle into a multitude of geometric figures: the square, the rectangle and the pyramid, through the reconversion of the diagonals into axes of symmetry, under the impetus of the arithmetical triplet.

8-) Crossed, convex and concave quadrilaterals

We came up with the idea of referring to 'diagonometry' as the branch of mathematics, or more precisely geometry, that focuses on the study of the behaviour of diagonals within a given geometric shape. By using their posture as an axis of symmetry, we move from one geometric figure to another, which is how we moved from the triangle to the square. However, the focus will be on the rectangle as a geometric component of the hexagram. In fact, as with crossed triangles, we also discover that rectangles are also crossed. But the most important thing to note is that the crossed rectangles are actually inscribed within a virtual rectangle, according to the very position of the diagonals. This is why we have decided to look at crossed, convex and concave quadrilaterals. It is judicious rightly to review our hexagram below:

Fig: the hexagram

After taking a closer look at this geometric shape, taking the whole plane into account, we discover that two rectangles intersect within another. The latter occupies a horizontal or normal position, while the crosses relate to its diagonals. In this first situation, they are therefore inside the ordinary quadrilateral. Studies of the internal position of the diagonals within the quadrilateral have shown that, if the four points ABCD are aligned so that (AB)=(CD) and (AD)=(BC) symbolise the lengths and widths respectively, then (AC) and (BD) constitute the digonals of this quadrilateral. Hence, the quadrilateral is said to be convex due to the action of the diagonals:

"convex A: the 2 diagonals are on the inside".

We therefore speak of a convex quadrilateral when its diagonals are located within it, while intersecting at their midpoint, which is the centre of symmetry. It's important to emphasise in passing that it's actually the core of the geometric figure, including the diagonals, that favours this designation and not its external configuration. The three rectangles of the hexagram thus constitute a convex quadrilateral.

In the second configuration, we now need to make one small change, through the straight relationship between the three rectangles. The main focus will be on the diagonals, before the lengths and widths of the virtual rectangle. This suggests that the diagonals will first be extracted from the rectangular grid for custom analysis, before being transcribed later. It is therefore appropriate to consider the diagonals (AC) and (BD), before later considering the dimensions (AB), (CD), (AD) and (BC), so we say that the diagonals are on the outside. Note that:

"crosses A: the 2 diagonals are on the outside".

In this configuration, the sides (AB) and (CD), then the sides (AD) and (BC), are complementary to the diagonals (AC) and (BD). The outer side of the quadrilateral is influenced by its inner side, which suggests that it is the diagonals that ultimately animate the geometric shape. Hence, the diagonals (AC) and (BD) form the primary shape of the quadrilateral.

The third configuration consists of reconsidering the rectangle from the first situation (crosses), recognisable by the sides (AB)=(CD), (AD)=(BC), and the

132

diagonals (AC) and (BD). Side (AD) will then be modified by becoming a pointed angle inside the rectangle, giving us (AB), (BC), (AD) and (CD). We can see that the diagonal (BD) will be inside the quadrilateral by tracing the diagonal (AC), which will be outside this geometric figure. So we're talking about concave quadrilaterals:

"concave A: one diagonal is on the inside, the other is on the outside".

The above shows that the diagonals play their primary role within the quadrilateral, but they can also define its side. This third configuration is consistent with the case of the traditional Ngoma instrument or Sacred Harp, whose eighth (8th) string symbolises the base of the triangle, which in fact constitutes the main diagonal of the 9 cm long square. For the diagonal to play the role of the side of a geometric figure, it must first be similar to a triangle. In reality, points ABC, without taking point D into account, relate to a triangle. In this case, side (AC) forms the base of the triangle in question, while sides (AB) and (BC) join at its vertex.

"The transformation of a quadrilateral into a triangle is accompanied by the conversion of one of its diagonals into a triangular base. This is the g'om'trique ddorem pioneered by the traditional Ngombi instrument, thanks to the arrangement of the stringophones that make it up".

Diagonometry, then, is the study of the behaviour of the diagonals of a quadrilateral, favouring its transformation into a trilateral figure, through the systematic exclusion or otherwise of one of the diagonals, while the other now forms the basis of the new geometric figure obtained.

"Diagometry explains the transformation of a quadrilateral into a trilateral figure by the action of the diagonals."

The Menorah, the Hanukkiah, the Mvet Ekang and the Sacred Harp are all concave constructions, because the diagonal is used as the triangular base of their different geometric frameworks. It is therefore possible to speak of a 'concave triangle' to mean that it is originally drawn from a quadrilateral by the action of the diagonals. The presence of the diagonals within the square and rectangle reflects the idea that, in one way or another, the quadrilaterals remain linked to the trilateral figure. The hexagram is therefore a practical perception of this practical unity.

9-) Calculating values

Plucked-string musical instruments and candlesticks generally have a triangle as their main framework, but by means of the diagonal, which is actually its base, we have arrived at a square. This quadrilateral is in turn covered by four triangles meeting at the centre of the intersection of the diagonals.

The first thing to note is that the length of a given triangle is contained within a number that is one unit greater than it. For the first trilateral figure presents just a parallel and triplet perception of numbers, whereas the second involves one more unit on its vertex giving rise to perpendicularity. It is a quadruple view of the figures or numbers in the plane.

Secondly, the volume of geometric shapes becomes more imposing the larger the measurement of the length of the side of the square.

These two observations have made it possible to draw up a table of correspondences between the lengths of the squares and the triangles they contain. The internal behaviour of the diagonals is also highlighted in this table. The table is made up of seven (7) columns:

the first from left to right refers to the different lengths of the edge,

the second column shows the falling value of the tiles,

the third reports an increase in eyesight,

the fourth reveals their sums,

the fifth relates to their products,

the sixth to the product of triangular values taken together,

finally,

the seventh to the number of squares in a single trilateral figure.

The evolution of arithmetic values from the square to the triangle

Geometric figures	Carre		Odd-numbered triplets			Triangle
Numbers	Internal Value 1(VI1) n-2	Internal value 2 (VI2) n-1	VI1+VI2	VI1XVI2	Multiplication by 2	Divion par 4
1	-	-			-	-
2	0	1	1	1	2	0,5
3	1	2	3	2	4	1
4	2	3	5	6	12	3
5	3	4	7	12	24	6
6	4	5	9	20	40	10
7	5	6	11	30	60	15
8	6	7	13	42	84	21
9	7	8	15	56	112	28
10	8	9	17	72	142	35,5
11	9	10	19	90	180	45
12	10	11	21	110	220	55
13	11	12	23	132	264	66
14	12	13	25	156	312	78
15	13	14	27	182	364	91
16	14	15	29	210	420	105
17	15	16	31	240	480	120
18	16	17	33	272	544	136
19	17	18	35	306	612	153
20	18	19	37	342	684	172
21	19	20	39	380	760	190
22	20	21	41	420	840	210

23	21	22	43	462	924	231
24	22	23	45	506	1012	253
25	23	24	47	552	1104	276
26	24	25	49	600	1200	300
27	25	26	51	650	1300	325
28	26	27	53	702	1404	351
29	27	28	55	756	1512	378

Sources: N'NANG EBANE Sosthene Tresor

The table above shows that the odd numbers are obtained by adding up the values inside the square according to specific lengths. They define an evolution in zigzag of the whole of the diagonals of the square, that is to say the image a "water undulates". These odd numbers are used to construct the rectangle in which the hexagram is inscribed. It is advisable to carry its glance on the column noted VI1+VI2.

The relationship between the square and the rectangle is also clearly established in terms of the distribution of numerical values according to the lengths of the given square. If the ordinary square is 8 cm long, then the vertical square is 9 cm long. In fact, the triplet of the number is 8+1+8=17 and 8+9+8=25, so the number 9 indicates the length of the vertical tile.

Obviously, there is also a close relationship between the length of the side of the square and the number of line segments parallel to the diagonal. If the side of the rhombus is 7 cm long, then it contains 10 line segments parallel to the diagonal. There are 11 parallel straight lines counted in one direction only, either from bottom to top or from top to bottom, 22 straight lines in both directions, and a further 22 in the other diagonal direction, the sum of which is 4x11=44.

The number 7 is always associated with the number 11, which adds up to 22, the symbol for the letters of the Hebrew alphabet. The combined luminous array of the Menorah and the Hanukkiah stagnates twice at the number 11, demonstrating the pervasive nature of the number 7 compared to the number 9. The former tends to make the latter less expressive. The number 7 is a singular perception of Hebraic life, centred around the geometric symbol of the pyramid.

The main diagonal of the square symbolises the number that is a tiny unit less than the length of the square. So, if the sides are each 10 cm long, the main diagonal is the 9th parallel of the plane. The Sacred Harp or Ngombi triangle is drawn from a square measuring 9 cm on each side, giving it 8 cordophones.

Other than the odd numbers taken as numerical triplets, the diagonal, which is the axis of symmetry within the square in relation to the line segments that surround it, is also arranged in triplets. If the side length of a rhombus is 9 cm, then its corresponding ascending diagonometric triplet is noted: 7+8+7=22. On the other hand, its descending diagonometric triplet is noted as 7+1+7=15. It is also possible to have an identical or stagnant diagonometric triplet with the following notation:

8+8+8. Hence the respective mathematical formulae, T(n)=n+(n+1)+n, T(n)=n+1+n, and T(n)=n+n+n equals n 3.

In reality, the numerical triplet of stagnation imposes the destruction of the square to consider neither more nor less than the triangle. Remembering the Ngoma triangle, which is half the square and 9 cm long, we can see that it has 8 upper perforations, 8 cordophones in the centre and 8 lower perforations, i.e. a purely parallel diagonometric grip. So this triplet corresponds to the formula: T(n)=n+n+n equals T(n) 3 n. So T(8)=8+8+8 is equivalent to T(8)= 3x8=24.

The numerical triplets appear on the square because the triangles are covered on the square, according to the insertion of cordophones within the geometric shapes, which is an ancestral African practice. "The need to make the pyramid less heavy is a desire that has long been expressed through this ingenious know-how".

While the first table presents only a bipartite distribution of values within the square and the triangle, the second emphasises a tripartite division, the rectangle, the square, then the triangle, including the hexagram in a globalising geometric figure. The table is composed of seven (7) columns:

the first (1st) from left to right refers to the different digits and natural numbers

the second (2nd) refers to descending numerical triplets, the sum of which relates to the number itself

the third (3rd) reports on the increasing view of numerical triplets (ascendants),

the fourth (4th) reveals their sums, those of the ascending triplets

the fifth (5th) relates to the totals of the tiles in the square according to the specified numerical values

finally,

the sixth year (6eme) to the number of squares in a trilateral figure individually.

It should be noted, however, that the number of squares in a crossed rectangle is the sum of the two triangles opposite the square, plus the contents of the square.

Variation in arithmetic values for rectangles, squares and triangles

		Rectangle			Carre	Triangles
Figures and numbers	Descending triplets T(n)=n+1+n	Ascending Triplets T(n)=n+(n+1)+n		Total	Number of tiles in the square V(n)-1=VI VI xVI	Number of squares in triangles
1	0-1-0	1		1	0	0
3	1-1-1	1-2-1		4	1	1
5	2-1-2	2+3+2		7	4	6
7	3-1-3	3+4+3		10	9	15
9	4-1-4	4+5+4		13	16	28
11	5-1-5	5+6+5		16	25	45
13	6-1-6	6+7+6		19	36	66
15	7-1-7	7+8+7		22	49	91
17	8-1-8	8+9+8		25	64	120
19	9-1-9	9+10+9		28	81	153

21	10-1-10	10+11+10	31	100	190
23	11-1-11	11+12+11	34	121	231
25	12-1-12	12+13+12	37	144	276
27	13-1-13	13+14+13	40	169	325
29	14-1-14	14+15+14	43	196	378
31	15-1-15	15+16+15	46	225	435
33	16-1-16	16+17+16	49	256	496
35	17-1-17	17+18+17	52	289	561
37	18-1-18	18+19+18	55	324	630
39	19-1-19	19+20+19	58	361	703
41	20-1-20	20+21+20	61	400	780
43	21-1-21	21+22+21	64	441	861
45	22-1-22	22+23+22	67	484	948
47	23-1-23	23+24+23	70	529	1035
49	24-1-24	24+25+24	73	576	1128
51	25-1-25	25+26+25	76	625	1225
53	26-1-26	26+27+26	79	676	1326
55	27-1-27	27+28+27	82	729	1431
57	28-1-28	28+29+28	85	784	1540
59	29-1-29	29+30+29	88	841	1653

Source: NNANG EBANE Sosthene Tresor

- Properties 1: Odd numbers and odd terminals

If the double digit in the ascending T(n) of a given number is odd and its own terminals in the descending T(n) are also odd, then the corresponding mathematical formula is: (11 2)-VC(n), where VC(n) represents the numerical value of the tiles in the vertical square.

Example 1:

T(n)=n+(n+1)+n equals T(3)=3+4+3=10

T(n)=n+1+n equals T (3)=3+1+3=7

Vc (n)=9

The number 9 indicates the number of squares in the vertical tile.

The number 3 has a negative triplet with its own good appearing doubly 111 (1 first and 1 last), by addition 1+1=2.

Vt(n)=(2*n)+VC(n) numerical application Vt(3)=(2*3)+9=6+9=15.

Or, Vt(n)=n+(n*n)+n equals vt(3)=3+(3*3)+3=3+9+3=15

The number 15 refers to the number of squares in the crossed triangles.

The number 7 (3-1-3)=(6x7)+49=42+49=91

Example 2:

The number 15 (7-1-7): (2x7)x15+225=(14x15)+225=210+225=435

The number 23 (11-1-11)=(2x11)x23+529=(22x23)=506+529=1035

The number 27 (13-1-13)=(26x27)+729=702+729=1431
The number 19 (9-1-9): (18x19)+361=342+361=703
* Properties 2: Odd numbers and even terminals
* Odd ascending triplet and even descending triplet

If the double digit in the ascending T(n) of a given number is odd and its own terminals in the descending T(n) are even, then the corresponding mathematical formula is: (nx4)+VC(n), where VC(n) represents the numerical value of the tiles in the square.

Example 1:

T(n)=n+(n+1)+n equals T(4)=4+5+4=13

T(n)=n+1+n equals T (4)=4+1+4=9

Vc (n)= (n-1)x(n-1) equals Vc(n)=(9-1)x(9-1)=8x8=64

The number 9 has a positive triplet with the good ones appearing twice 414(4 first and 4 last), by addition 4+4=8.

Vt(n)=(4xn)+VC(n) numerical application VT(9)=(8x9)+81=72+81 = 153.

Example 2:

The number 5 (2-1-2): (2x2)=4 and (4x5)+25=20+25=45

The number 13 (6-1-6): (2x6)=12 and

(12x13)+156=156+169=325

* Even ascension triplet (bounds) and even regression triplet (bounds).

If the double digit in the ascending T(n) of a given number is even and its own bounds in the descending T(n) are even, then the corresponding mathematical formula is: (V(n)-1)x n)+VC(n), where VC(n) represents the numerical value of the tiles in the square.

Example 1:

T(n)=n+(n+1)+n equals T(10)=10+11+10=31

T(n)=n+1+n equals T (10)=10+1+10=21

Vc (n)=100

The number 10 has a positive triplet with the good ones appearing twice 424(4 first and 4 last), by addition 4+4=8

Vt(n)=(V(n)-1)xn+VC(n) numerical application VT(10)=(10-1)x 10+100 = (9x10)+100=100+90=190

Example 2:

The number 8: (8-1)x8+64=(7x8)+64=64+56=120

The number 14: (14-1)x14+196=(13x14)+196=182+196=378

The number 18: (18-1)x18+324=(17x18)+324=306+324=630

The number 22: (21x22)+484=462+484=948

10-) Triangular and square content

It has been shown throughout the preceding analysis that the triangles of the Mvet instrument and those of the Menorah and Hanukkiah, are in fact made up of a vertical square supported by two opposing isosceles triangles, whose image may refer to the sun rising between two hills whose summits are sharp' s. This imagery is of course related to the thought expressed by the name of the object of study, because the name Mvet actually refers to the fact of detaching oneself from the earth and reaching for the heavens, and therefore to elevation. This notion also reveals our ability to perceive and interpret the different images inscribed within it, because it makes an object that we think of as static, mobile. Elevation is not just a matter of perceiving an object that has left the earth to take refuge in the heavens, but also, and above all, of manipulating it through a methodical demonstration aimed at deciphering the codes that make it up. In this analysis, we need to determine the degree of stagnation of a triangle, not using the luminous table, but from the numerical triplet from which it draws its essence. This will allow us to determine the number of squares in the square and the triangles that make up its base. Let's consider the number 7, which is the third triplet.

We know that

T(n)=n+1+n

T(3)=3+1+3

T(3)=7

which becomes

T(n)=n+(1+n)+n

T(3)=3+(1+3)+3

=3+4+3

T(3)=10.

From this result, we take the number in the centre, i.e. 4, and subtract one unit (1) from it, then add the result to the square, i.e. nxn. It should be noted, however, that the formula is adapted according to the configuration of the triplet.

We pose:

T(n)=n-1+(n-1) 2+n-1^A

T(4)=(4-1)+(4—1)A2+(4—i)

=3+(3) 2+3^n

=3+(3x3)+3

=3+9+3

T(4)=15.

An image of the pyramid seen from this angle is of vital importance:

Fig: a shattered pyramid

The figure above effectively shows a square detached from the triangles and reaching for the heavens, like a rocket taking off to conquer space. The square undoubtedly has the largest number of squares, 9 against 3 on each side. This is the image buried within the Mvet Ekang instrument and the Jewish candlesticks, which can clearly attest to the fact that these peoples belonged to the Pharaonic kingdom, including the Punu people.

The vertical square on the triangles has 9 squares, while the triangles on the left and right each have 3 squares. The number 15 refers to the degree of stagnation of the Menorah or 7-branched candlestick. It is also presented as a triplet of the number 5, i.e. $5+5+5=3\times5=15$. The complete triangular numerical condensation is $2+4+5+5+4+2=27$.

As for the Hanukkiah or the 9-branch candlestick, we put it down:

$T(n)=n+1+n$

$T(4)=4+1+4$

$T(4)=9$

which subsequently becomes

$T(n)=n+(1+n)+n$

$T(4)=4+(1+4)+4$

$=4+5+4$

$T(4)=13$.

Then the following formula applies:

$T(2)=2(n-2)+(n-1)^2+2(n-1)$

$=2(5-2)+(5-1)^2+2(5-2)$

$=2(3)+(4)^2+2(3)$

$=6+16+6$

$T(2)=28$.

The result obtained below can effectively take shape within the pyramid sketch above:

Fig: a shattered pyramid

Source: NNANG EBANE Sosthene Tresor

Decidedly, the edge is always and always of greater value. The question of a collective study of our civilisational memories therefore remains a definite priority, in order to ward off the standardisation of knowledge that has been excessively offered to us over the years. Every people has an authentic character that should never be forgotten.

The number 28 defines the stagnation of the 9-point candlestick, i.e. 7+7+7+7 = 4x7=28. The square is made up of 16 squares, while the triangles each have 6 squares.

✓The number 15

We pose:

7+1+7=15 equals 7+8+7

then,

$$T(3)=3(n-1)+(n-1)^2+3(n-1)$$

$$=3(8-1)+(8-1)^2+3(8-1)$$

$$=3(7)+(7)^3+3(7)$$

$$T(3)=21+49+21.$$

$$T(3)=91$$

✓The number 17

8+1+8=17 equals 8+9+8

then,

$$T(4)=4(8-2)+2(n-2)^2+4(n-2)$$

$$=4(8-2)+2(8-2)^2+4(9-2)$$

$$=4(6)+2(6)^2+4(6)$$

$$T=24+(2\times36)+24$$

$$=24+72+24$$

$$T(4)=120$$

In the light of the above, it should be noted that the formula used for the number 7 is doubly perceptible within that of the number 15. In simple terms, the triplet 3+1+3=7 is counted twice within the number 15 respectively, (7=3+1+3)+1+(3+1+3=7) equals 7+1+7=15, and this number also constitutes the stagnant value of the number in question. In the same vein, the triplet of the number 9, 4+1+4, is also present in the number 17, because the former is the degree of angular growth of the latter. So 8+(4+1+4)+8 equals 8+9+8.

The configuration of digits and numbers in the triplet, uniting the square with the triangle, always reveals the numbers 22 and 26 repeatedly. For example, 7+15=22 and 9+17=26, or 9+13=22. The number 22 defines the letters of the Hebraic alphabet, while the number 26, again for the Jews, refers to the name Yahwhe (God). In the light of the above, these two numbers can obviously be considered as authentic numerical codes, revealing the presence of the divine within the religious instruments Mvet Ekang or Harp-Citar, the Ngombi or Sacred Harp, the Menorah or 7-branched Candelabra, the Hanukkiah or 9-branched Candelabra, all related to the pyramid. This mathematical revelation effectively demonstrates that the so-called Bantu peoples and the Jews do indeed share a common destiny, not to say a common history.

11-) Numerical triplets and their bounds

The evolution from numeric triplets to their bounds refers to the fact that it is possible to adapt a mathematical formula to each digit or number, depending on whether its symmetrical bounds are even or odd. A symmetrical boundary is taken to mean the digits and numbers located at the extremities of the digit 1 taken to be the centre of the axis, serving as the image of one or other of them in relation to the centre. These mathematical formulae allow us to determine the number of squares contained in a complete equilateral triangle, divided into two small rectangular isosceles triangles and one square, when a pyramid is viewed from the front and from the inside. Let's take a closer look at the following examples:

✓The number 3

Triplet note 1+1+1

T=(n-1)+(n-1)^2+(n-1)

=(2-1)+(2-1)^2+(2-1)

=1+(1)^2+1

=1+1+1

T=3

The number **3 is the first (1st)** numerical triplet with purely negative symmetric terminals, hence it includes an adapted mathematical formula which is naturally odd, denoted (n-1).
✓The number 5
Triplet note 2+1+2

T=(n-2)+(n-1)^2+(n-2)

=(3-1)+(3-1)^2+(3-1)

=2+(2)^2+2

=2+4+2

T=6

The number **5 is the second (2nd)** numeric triplet, but contains positive symmetric terminals, hence the mathematical formula (n-2).
✓The number 7
Triplet note 3+1+3
T=(n-1)+(n-1)^2+(n-1)

=(4-1)+(4-1)^2+(4-1)

=3+(3)^2+3

=3+9+3

T=15

The number **7 constitutes the third (3rd)** numerical triplet and has purely negative symmetrical limits, i.e. 3 against 3, which is why it is adapted to the mathematical formula (n-1). It is fair to say that the first numeric triplet, 3, is revealed within the third, i.e. the number 7.
✓The number 9
Triplet note 4+1+4

T= 2(n-2)+(n-1)^2+ 2(n-2)

=2(5-2)+(5-1)^2+(5-2)

=2(3)+(4)^2+2(3)

=6+16+6

T=28

The number **9 is the fourth (4th)** numeric triplet and the second (2nd) to include positive symmetric terminals. This is how the formula for the first numeric triplet, comprising the first positive symmetrical terminals, is completed by the digit 2 as a factor noted 2(n-2).
✓The number 11
Triplet note 5+1+5

T=2(n-1)+(n-1)^2+2(n-1)

=2(6-1)+(6-1)^2+(6-1)

=2(5)+(5)^2+2(5)

=10+25+10

T=45

The number **11 is the fifth (5th)** numerical triplet, and contains negative symmetric bounds 5 against 5. However, the number 5 contains positive symmetric bounds, hence the mathematical formula 2(n-1).
✓The number 13
Triplet note 6+1+6

T= 3(n-2)+(n-1)^2+ 3(n-2)

=3(7-2)+(7-1)^2+3(7-2)

=3(5)+(6)^2+3(5)

=15+36+15

T=66

The number **13 is the sixth (6)** numerical triplet and has possible symmetrical bounds. However, its bounds are double those of the first numerical triplet, which is why the mathematical formula 3(n-2) is adapted to it. Bear in mind that the number 6 is half the number 12.
✓The number 15

Triplet note 7+1+7

$$T=3(n-1)+(n-1)^2+3(n-1)$$

$$=3(8-1)+(8-1)^2+(8-1)$$

$$=3(7)+(7)^2+3(7)$$

$$=21+49+21$$

$$T=91$$

The number **15 is the seventh (7)** numerical triplet and comprises symmetrical negative 7 against 7, which is why the formula 3(n-1) is the most suitable. It is important to remember that the number 15, i.e. 5+5+5 or 3*5, is the stagnant numerical value of the numerical triplet 3+1+3 within the 7-point candlestick.
√The number 17
Triplet note 8+1+8

$$T=4(n-2)+(n-1)^2+4(n-2)$$

$$= 4(9-2)+(9-1)^2+4(9-2)$$

$$=4(7)+(8)^2+4(7)$$

$$=28+64+28$$

$$T=120$$

The number **17 is the eighth (8th)** numerical triplet and includes possible symmetrical terminals. However, the number 8 is double the number 4, hence the mathematical formula 4(n-2).

The passage from the number 17 to the triplet 8-1-8 corresponds to a passage from the closed pyramid seen from the sky to the so-called open pyramid, i.e. the unification of the square at the centre of the four isolated triangles. The triplet highlights the idea that two elements of the same nature can only be united by a foreign or neutral energy. The sky and the earth are united by the light of the sun, the moon, the stars and so on. The notion of central and othogonal symmetry actually brings to the fore the idea of coming together, of unity in a common cause, around an individual who knows how to lead the rest of the group. The number 1 in the centre is a symbol of unity, of the common thread, of coming together around a common ideal, like the branch of a palm tree, a stream flowing into another, the division of a branch into two arms, etc,.

Triplets 3 and 7 refer to the number 12.

We have: 111+313=444 equals 4+4+4=3x4=12.

Geometry and algebra are combined in Jewish candlesticks, as two inseparable branches of mathematics, around the different configurations of the pyramid. It is a synthetic perception of the square, triangle and rectangle, through the conversion of the diagonals into axes of symmetry and the lengths of triangles and rectangles.

VI-) THE CALENDAR

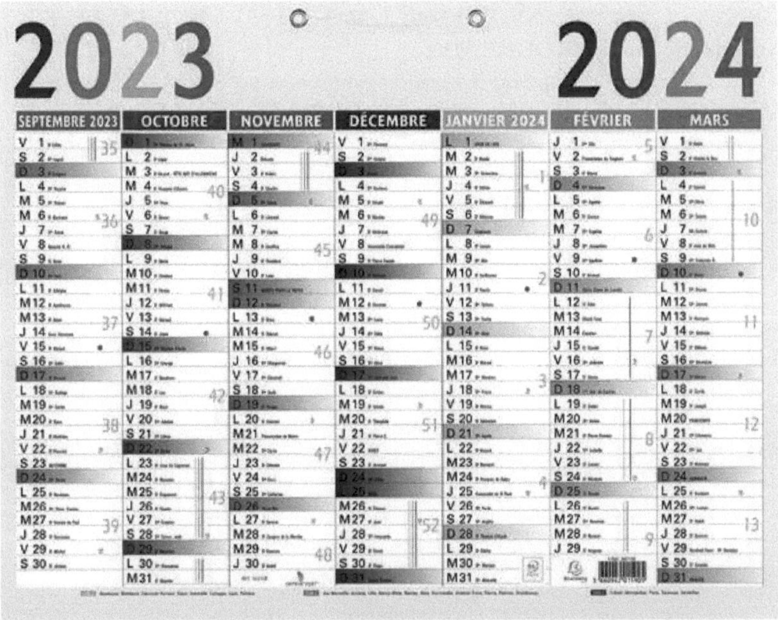

Source: www.amazon.fr/calendrier-multicomore-43c...

A
our late
classmate
ENVOLE Jessy Lionel (Tle A1)

(Lycee public Moise NKOGHE MVE)

1-) The number 7 and the days of the months

It is generally known that the current calendar has 12 months in a year, but not all months have the same number of days. Some months have 28 or 29 days, others 30, and still others 31. From the above, we can see that the months of the year have a fourfold distribution (4) if we consider the two variations of February, and a threefold distribution (3) if we consider one of the values of this same month over the course of the year. It is thus presented in a twofold way: 3+4=7. The figure is therefore the result of the sum of an odd number and an even number, which differ by just one tiny unit. We count 28, 29, 30 and 31, then 28, 30, 31 or 29, 30, 31. Everything seems to indicate that the number 4, symbolised by the numbers 28, 29, 30 and 31, is eclipsed to make way for the number 3, whose numerical correspondents are 28, 30 and 31 or 29, 30 and 31.

The number 4 has only one (1) possible designation, while the number 3 has two (2), i.e. with the variations in the month of February. It would therefore be possible to calculate the total number relative to the frequency of their appearance, i.e. (2 3) + 4 6+4 10. It was demonstrated at length above that the number 7 comprises two arithmetical triplets, 3+1+3=7 and 3+4+3=10 respectively. This is the second triplet, known as the ascent triplet, which in fact accounts for the result obtained. The number 3, doubly perceptible within the calendar, is also doubly located around the number 4, if we consider the second triplet. In contrast, the number in the centre remains immobile. To illustrate the above in figures, let's say: (28+30+31)+(28+29+30+31)+(29+30+31)=89+118+90. The evolution of the numbers in the set defines an object, capable of starting from a lower level to reach the top, before falling later. It would therefore be wise to dwell on texts relating to the origin of the calendar:

"The construction of calendars is complex and varies according to culture. Some have chosen to follow the moon, others the sun or sometimes both. The construction of calendars calls on ma^matics, cultural aspects, and demonstrate ingëniosité and adjustment."

The notion of triplicity may refer not only to the idea of a tripartite vision of an object, but also to its capacity to undergo multiple mutations over the course of a given period. The moon, like the sun, is in perpetual motion, and experiences periods of very intense and less intense luminosity. Hence the numbers 89 and 90, which clearly show this from a purely arithmetical point of view. As far as the sun is concerned, its position at the zenith tends to define it as brighter than its rising and setting. So the number 3 would designate the sunrise, 4 the zenith, and the second 3 the sunset, all in a triangular shape. The mention of mathematics in this passage explaining the calendar suggests that the elements of the universe have been transposed into the realm of human thought, with the aim of understanding and manipulating them in order to better organise our social lives. There are many phases to the moon, but it is only fair to offer a practical view of one of them:

"The New Moon: During this phase, the moon lies between the earth and the sun,

which means that the side we see is not ёCmmёв, that the moon appears almost invisible in the night sky."

According to this passage, the position of the moon in the centre of the earth and the sun corresponds without a shadow of a doubt to our numerical triplet taken as an illustration. The sun and the earth symbolise the numbers 3 on the far right and far left, while the moon represents the number 4 in the centre. The transition from the regression triplet 3+1+3=7 to the ascension triplet 3+(1+3)+3=3+4+3=10 suggests that the central element is somehow influenced by the elements around it. It is said that the moon produces no heat of its own, but that it is the reflection of the sun that provides the light we see through it.

Let's go back to the consecutive digits 3 and 4, which make up the number 7. It should be remembered that, in the context of the calendar, the former can be seen two (2) times, while the latter offers only one (1) possibility. The triplet is not only no longer reflected in the arithmetical components of the number 7, but also in the frequency with which they appear. In this case, the number 3 will appear exactly 3 times, while the number 4 will appear one (1) time only, so we go from (3+3)+4=6+4=10 to (3+3+3)+4=4+9=11. This new number refers not only to the number of diagonals in a 7cm square, but also to the number of intervals in the number 12. Although the number 12 denotes the degree of stagnation of the number 8, it is also associated with the number 7. The perception of this mythical number as the central axis of the calendar will be the subject of the next point.

2-) The central axis of the calendar (24X)

We all know that today's calendar is made up of 7 months, each lasting 31 days. This fact is in line with the expression of stagnation that we discussed earlier. The other 4 months of 30 days fit in with the above. If there is stagnation, this means that there is also the possibility of both ascension and regression. The month of February alone counts either 28 or 29 days, moving from the smaller value to the larger one by a small unit. The numerical triplet of the ascending number 7 has revealed that the number 4 constitutes its arithmetical centre of symmetry, and then by extension that the number 28 relates to the triangular stagnant value of the number 9. However, we can see that the numbers 7 and 4 always follow each other, based on the consecutive ranking of the months with the greatest number of days.

We now need to draw a horizontal axis to which we first associate the 31-day months. We will then locate all the months with less than 31 days at the lower centres of each of their intervals. In other words, the month of February and the 4 months with 30 days. The aim is to reconcile the number 7 with the number 12. Each dot represents a month of 31 days:

1*   2*   3*   4*   5*   6*   7*

Next, let's locate the months under 31 in each of their lower centres.

150

1*   2*   3*   4*   5*   6*   7*

1•   2•   3•   4•   5•

In view of the above, the 12 months of the year are effectively built around the calendar axis 7. There are 7 central months and 5 lower months. The year is read from the central number 1 to the lower number 1, then from the central number 2 down to the lower number 3, and so on to the end. The numbers 4 and 5 have no lower centre, because they symbolise the months of July and August respectively, each of which has 31 days, hence the 'stagnation'. From this point of view, the twelfth (12) month merges with the number 7. It would not be wrong to present this perception in a more obvious way:

The digital time structure

1.   2   3.   4.   5.   6.   7.

      *3   *5   *   *   *10   *12
                  7   8

2.   4   6   9   11

Source: NNANG EBANE Sosthene Tresor

The figures in black refer to the seven months of the year, each with 31 days, those in blue designate the months with less than 31 days, and the red ones refer to the different undulatory values, supported by the figures and numbers in blue. The wave variations are smaller than the different numbers on the central axis. The curve corresponding to this structure of arithmetical figures and numbers is as follows:

The time curve

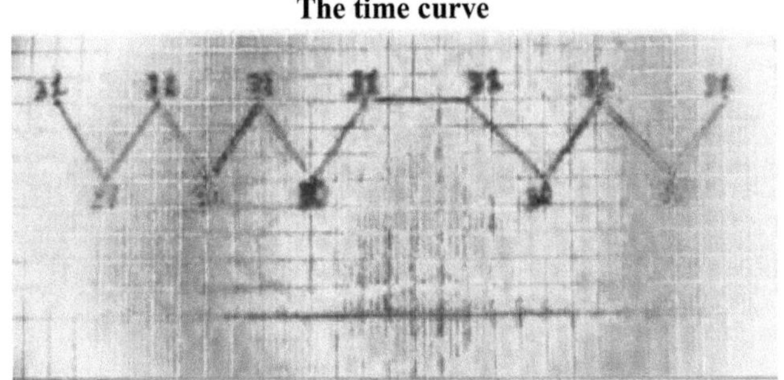

Source: NNANG EBANE Sosthene Tresor

The curve shows undulating movements from January to July, then from August to December. Stagnation only occurs between July and August. Everything suggests that there is no evolution that does not experience a period of stagnation, or quite simply of rest. The question of whether there is a happy medium in the choice of life is thus perfectly answered.

We can see that the number 7 is associated with the number 12, proving that the latter symbolises the universal axis. We can also see that the number 7 in red, symbolising the month of July, is linked to the number 4 at the centre, just as the number 8 in red is associated with the number 5, and the number 4 is also linked to the number 9. The number 4 represents the arithmetical centre of symmetry of the ascending triplet of the number 7, and at the same time, the symmetrical boundary of the number 9. The number 5, on the other hand, represents the triangular stagnant value of the number 8, and the centre of symmetry of the number 9. We discover that the data from the squares and luminous arrays of the numbers 7, 8 and 9 are included within the calendar.

It was shown above that the triplet 789 relates to the number 24, i.e. 7+8+9=24. Now let's see how this number appears diagonally across the centre of the arithmetic axis:

The diagonometric core (X) of the axis

4   5
*   *
7   8

Source: NNANG EBANE Sosthene Tresor

We can see that (5+7)+(7+5)=12+12=24 and (4+8)+(8+4)=12+12=24. The centre of the 7 calendar axis shows the number 24 appearing 4 times. Remembering that the square is 8 cm long, we can see that its diagonals stagnate exactly 4 times at 12, so the number 8 is the diagonometric centre of the 7 axis.

We also see that (4+7)+(5+8)=11+13=24 and (7+4)+(8+5)=11+13=24, and (4+5)+(7+8)=9+15=24 and (5+4)+ (8+7)=9+15=24. So the lengths of the digits (‖) and (=), like the digonals (x), all relate to the number 24. So the numbers 4, 5, 7 and 8 symbolise the square. The calendar axis (7) has a square at its centre.

3-) The central axis of the calendar (25X)

The first arithmetical time structure has two fundamental aspects: undulation and stagnation. The latter is effectively caught between two undulatory variations, the first running from January to July, and the second starting in August and ending in December. This gives rise to a tripartite arrangement of the calendar, with stagnation corresponding to the number 24.

However, we need to sweep the central stagnation, so that we have wave variations all along the axis, in order to discover whether it is possible to obtain another stagnant value? The following numerical diagram illustrates this perfectly:

The digital time structure 2

<pre>
1.   2   3.   4.   5.   6.   7.
     *3   *5   *   *   *11   *13
                   7   9
   2.   4   6   8   10   12
</pre>

152

This structure offers a permanent wave-like vision. The number 8 is no longer completely adjacent to the number 4, as it has become the lower centre of numbers 4 and 5, accompanied by the rise of the number 9 alongside the number 5. In this configuration, the ascension of the numbers 7 and 9 reaches saturation point. In fact, the number 4 symbolises the numerical symmetrical centre of the number 7, just as the number 5 does for the number 9. This is how the X diagonals take on another special configuration:

The diagonometric core (X) of the axis

$$
\begin{array}{cc}
4 & 5 \\
* & * \\
7 & 9
\end{array}
$$

We see that $(5+7)+(7+5)=12+12=24$ and $(4+9)+(9+4)=13+13=26$, and 24 is different from 26, so one of the diagonals of the square is longer than the other. It therefore grows by two units more than its counterpart. The number pair (4;9) refers to the longest diagonal, while the number pair (5;7) refers to the shortest. This suggests a diagonometric enlargement of the square, i.e. a wider view of the quadrilateral. It is true that the square becomes wider through the right angles, but the symmetrical centre of the diagonals does not lose its right angle.

The first configuration of the axis shows stagnation, while the second shows continuous undulation. The numbers 24 and 25 thus appear consecutively across the two time axes. While the first number indicates diagonals of equal length, the second is distinguished by one diagonal being longer than the other. The two configurations taken together define an element capable of knowing both the ascent, the pairs (5;7) and (4;9), and the regression, the pairs (4;8) and (5;7). In practice, only February has this double variation.

A primordial idea emerges from this observation, relating to the presence of a life-giving matter within another to be brought to life. It must be admitted that within every moving or immobile body there lives a being that gives life to the latter, whether or not it is perceptible to the naked eye. The limit of science is clearly established by the systematic rejection of the spiritual order. The experimental perception of science is a reduced view of science. It amounts to saying that the world of intelligences is governed by divinities, who impose the conduct of science on humanity. According to the African understanding, science imposes an unconditional recognition of the use of knowledge as a loan. Initiation leads humanity to know how to silence its carnal desires, in order to act according to the strict vision of science. The first axis can be used to define a human being who knows how to contain himself, while the second refers to the animal nature that sometimes leads him to behave in an unworthy manner. The state of nature rules out any human knowledge of the object, which in reality is the object of science.

Pretending to study man while excluding his creator is a pure and simple illusion. He is only the manifestation of the spirit. The diagonals are the life-giving part of the square, while the sides are only affected by them, hence the term diagonometry. The calendar is not just about knowing the date of the day, the months of the year, but also about connecting humanity to its universe: to planetary phenomena. The calendar as a temporal object disappears, to reveal a global view of the workings of the solar system. The different phases of the moon, the tripatite journey of the sun, the alignment of the sun in relation to the moon and the earth, the life of the planets, and much more. The calendar then becomes a portal of escape or the very growth of human divinity. The observation is even beyond the limits set by the object, through its immaterial understanding.

4-) The diagonometric heart and the Sacred Harp

As an introduction to the observation relating to the mathematical essence of traditional African arts, it has been shown that the Sacred Harp has a triangular arrangement of cordophones: 8 upper perforations, 8 cordophones, and 8 lower perforations. If we analyse the calendar from a purely arithmetical point of view, we can see that the 7 axis is diagonometrically stagnant around the number 24. This number is thus buried in an identical triplet form: T(n)=n+n+n=nx3 equals T (8)=8+8+8=3x8=24. Let's consider the first square pair (4;5;7;8), to demonstrate this clearly:

The diagonometric core (X) of the axis

$$
\begin{array}{cc}
4 & 5 \\
* & * \\
7 & 8
\end{array}
$$

Source: NNANG EBANE Sosthene Tresor

Considering the triangular pair (4; 5; 7), we see that their sum is twice the number 8, i.e. 4+5+7=16, or 2^8=16, hence the triplet 8+8+8=3x8=24. Remembering that the square has 8 perforations or 8cm of sides, it has a triangular stagnation of 12, i.e. 12x4=48. So the diagonals of the square, whose sides are 8 cm long, form the symmetrical heart of the 7 axis of the calendar. In reality, the number 7 is the main diagonal of this quadrilateral. We can see that this triplet is taken from a square and not from a triangle, which means that the trilateral figure is a clear component of the square. Hence the terms "diagonometry" and "triangular concavity".

Let us now consider the second diagonometric core of axis 7 below:

The diagonometric core (X) of the axis

$$
\begin{array}{cc}
4 & 5 \\
* & * \\
7 & 9
\end{array}
$$

Source: NNANG EBANE Sosthene Tresor

The triangular triplet is still in place, but there is an increase of one diagonal unit from 8 to 9. If the triplet 457 corresponds to double the number 8 (2x8=16), the number 9 is an increase of one unit of this number 9=8+1, so we have 8+8+8+1. T (n)=n+n+n+1 equals T (8)=8+8+8+1=(3x8)+1=24+1=25. We remain here in the February mobility driven by the diagonometric core of axis 7. It's worth pointing out in passing that there are also 8 vices, in addition to the 888 triplet, to designate the diagonometric heart of the axis. It should be illustrated as follows:

The diagonometric core (24X) of the axis

8   8

*   *

8   8

Source: NNANG EBANE Sosthene Tresor

On the Sacred Harp the square is divided in two (2) by one of the main diagonals, so that we have 888, corresponding exactly to the 8 upper perforations, the 8 cordophones, and the 8 lower perforations, i.e. 3x8=24. The remaining 8 refers to the defects in the Sacred Harp's neck. In view of the above, there is not the slightest doubt that the ancestral Ngombi is part of the very fabric of time. It is a pure work of diagonometry.

The relationship between the number 9 and the number 8 is established in the sense that the latter is a numerical value internal to the former. But how can this relationship be made clear through the geometrical figures of the square and the triangle? The main idea that emerges is that one cannot be expressed without the other. So the "diagonometric triangle" or "concave triangle", whose sides are 8 cm long, can only be derived from the square which is 9 cm long. It is therefore a diagonometric and parallel take of the triangle within the square. The number 8 symbolises not only the main diagonal and the axis of symmetry, but also the base of the triangle. The eighth cordophone of the Sacred Harp, traditionally known as the Ngoma/Ngombi, forms the base of the triangle. The apex is not visible on the Ngombi, as it is a diagonal and parallel perception of the trilateral figure. The number 9 symbolises the square, while the number 8 represents the triangle. A simple rule emerges from the above: the diagonal and parallel triangle is the result of subtracting the length of the side of the square from a unit. In simple terms, we can only draw 8 parallel diagonometric and fractal line segments within a square whose sides are 9cm long.

The Sacred Harp's cordophones constitute its flexible part, whereas the neck and the rest of the instrument are not. What's more, their fractal, parallel and diagonal arrangement makes them a trilateral figure. The 8 cordophones actually symbolise the diagonals of the square in a triangular and fractal form. In the same vein, the figures arranged in squares and diagonals on the 7 axis of the calendar offer a threefold perception of the figure 8 that makes up their number. In other words, the changes in the calendar that can be observed throughout the month of February are celebrated by the sounding of the Sacred Harp, via the mobility of the cordophones, and therefore of the diagonals.

Flexibility and the trilateral figure combine to express mobility, ascent and regression. The triangular-shaped cordophones offer an enlivened perception of the triangular shape (animation geometry). When the instrument is played, the triangle moves away from the square towards the heavens. The music, the rhythm, the echo, being the work of the abstract, which launches a solemn appeal to the human soul to detach itself from the body in order to manifest itself freely, expresses a connection from the invisible to the invisible. It is an appeal to the state of nature of the object (matter) through its own destruction, which is not perceptible to the naked eye. The square, which is matter, can only contain the triangle, which is spirit, for a given period of time, before it escapes. When the instrument is left at rest, the triangle falls and merges back into the square, and the soul returns to its bodily envelope.

The month of February, made up of 28 days, relates to the year of self-closure, self-criticism. The most explicit corresponding image would relate to agriculture, through the planting of seeds. On the other hand, the 29th day would designate openness to other aspirations, which would correspond to the harvest period. The symbolism of the triangle with the square, through the ascent of the first on the second, is explained by the fact that we have realised that the human being is a work of the spirit, and that it is constantly evolving. In reality, this evolution is synonymous with the pursuit of a path, a path that was traced long before him. The invisible or abstract is both anterior and future to him, while matter is central to him. In this way, the triplet spirit-matter-spirit sums up human life, not only before birth, during life on earth, but also afterwards. The number 3 thus evokes the idea of a triple life.

The introduction of music is related to the fact that every universal creature can be identified by a particular sound. The crow of the cock is heard by all, but the tripatite journey of the sun across the sky is not audible by all, which is why a certain science is needed. Humanity, being the work of the spiritual, must connect with other entities of the same order, in order to understand life. It is not the only living being on earth, still less the only energy capable of materialising. Respect for the environment, strictly speaking, will help it to discover that there are other nations on earth, other forms of life, with which it must collaborate. It is therefore a voice that often leads the soul to leave the body, accompanied by a vision of things that are immaterial to the carnal man. The intoxicating song of a Cyrene in cartoons is a perfect illustration. Hearing and sight are two senses that merge to penetrate the spiritual universe.

5-) Wave variations and the Mvet Ekang

The Mvet Ekang instrument, like the Ngoma, is a plucked string instrument, according to an innocent view of these civilisational memories. For a more enigmatic vision, we need to reconsider the temporal axis 7 of the calendar, and add a completely different algebraic organisation. However, it would not be shocking to define them as permanent expressions of life, through echo or vibration, embodying the work of creation. Any object that we wish to create or materialise necessarily undergoes vibratory manipulation, in the hands of the creator, which makes these religious monuments timeless realities. According to the Mvet Ekang, time refers to the set of everyday events of which mankind is not the author, but a mere spectator or admirer. Time distances man from creative action, since he is himself the subject of creation. Time contains the idea of the elusive, of the inability to conceive, rather to undergo. Time is a destination taken against its grain, because unconscious at the moment when time chose to follow its course, but awakened later within a world that we struggle to understand for lack of time. Time is like quicksand, simply attracting everything that falls into it. Time is a vortex for every living thing within it. Present but powerless, that's what time reduces us to.

So as not to lose the thread of our ideas, we suggest returning to our subject, which is devoted to establishing the link between the wave values observable on the calendar and the morphology of the traditional Mvet Oyeng instrument. Let's consider our time axis below:

The digital time structure

1.	2	3.	4.	5.	6.	7.
	*3	*5	*7	*8	*10	*12
2.	4	6		9	11	

Source: NNANG EBANE Sosthene Tresor

The line of figures in blue will be completed above the 7 figures on the axis, in order to obtain both upper and lower ripples. It is important to remember that only the lower ripples are numerically represented below the stagnation period. The new configuration of axis 7 below is thus obtained:

The Wave Cycle

2.	4	6		9	11	
	*3	*5	*7	*8	*10	*12
1.	2	3.	4.	5.	6.	7.
	*3	*5	*7	*8	*10	*12
2.	4	6		9	11	

Source: NNANG EBANE Sosthene Tresor

The result is quite simply exceptional: the vertical triplets, the digits, and the numbers in blue, seem to relate to upper and lower triangles linked to a quadrilateral in the centre. As a practical example, the triplet 323 and the triplet 535, and the double number 4 above and below, define the union of two triangles with a quadrilateral. We can even add the numbers 1 and 4 on the horizontal plane, to obtain the exact image of a square associated with four triangles. This is a practical rendering of our open or split pyramid.

The break between July and August is palpable, as it harks back to the period of stagnation.

It is important, however, to present the second configuration of the time axis, for a better understanding of the Mvet Ekang instrument, according to the calendar. We need to add a wave value within the period of stagnation itself. The following is a good illustration of this:

The Wave Cycle

	2	4	6	8	10	12	
1	*3	*5	*7	*9	*11	*13	
	1.	2	3.	4.	5.	6.	7.
1	*3	*5	*7	*9	*11	*13	
	2	4	6	8	10	12	

Source: NNANG EBANE Sosthene Tresor

If you look closely at the wave cycle above, the number 1 is joined by two other numbers in blue at the upper and lower levels, because the tripet refers to the addition of three (3) elements. Excluding the period of stagnation, the number 9 takes the place of the number 8, which descends to the lower centre and rises to the upper centre of the interval. In total, there are six (6) quadrilaterals in vertical postures, rather than five (5) as in the first example.

Bear in mind that only the numbers in red and those in blue have both upper and lower triangular views. However, as far as the Mvet Ekang instrument is concerned, only the triangles pointing towards the heavens should be taken into account. In fact, the term 'Mvet' in the Fang language happily relates to the idea of ascent, greatness, summit, lordship and more. The civilisational memory is made up of a trestle pierced in a vertical position, triangular-shaped cordophones, horizontal branches and, just below them, half calabashes. This is now being reorganised according to the calendar:

✓The diagonals of the cmur of axis 7 are organised around the number 24, through the triplet 888, which becomes 8888 to signify that the triangle, although rising above the square, is also one of its coposants, and is also part of the Mvet Ekang. The bridge at the centre of the horizontal branch has 8 perforations and divides its base into two (2) equal parts of 8 perforations each. The 8 cordophones are clamped on either side of the horizontal branch, crossing the bridge in the centre. The Ekang Mvet, made up of eight (8) cordophones, refers to the number 24, which is the diagonal number of the 7 axis of the calendar.

✓ The horizontal branch, pierced in two equal parts, symbolises the temporal axis of the calendar, i.e. the 7 months of 31 days. It refers to stagnation and rigidity.

✓ The trestle drilled in a vertical position designates the height of the triangle, and the numerical symmetrical centre in relation to the notion of a triplet.

✓Cordophones are taken to be triangles of the wave cycle that tend only towards my heavens.

√The half calabashes under the horizontal branch can obviously define the lower undulatory movements as well as the stagnation period. They vary according to two orders in principle: the central calabash is the most voluminous, while the other two on either side are of the same size (1). All three (3) half-cups are identical; it is as if they were not there at all, or as if the instrument had no cups at all (2). The large half calabashes would therefore designate months of 31 days, the medium ones those of 30 days, and the very small ones the months of 28 and 29 days. It should rightly be pointed out that only the cordophones and the half calabashes convey the idea of ascension and regression, while the horizontal branch remains immobile.

In a certain theological dimension, this refers to the idea that life reigns and rules over everything that is matter. Matter is therefore in reality caught up in life on all sides. The image of an island can be used schematically, since the rising waters make it clearly non-existent. In the field of metallurgy, we understand that the solid is imperatively born of the liquid. So the liquid remains the solid's anteriority and its future, because by melting it, it becomes liquid again. The liquid-solid-liquid triplet relates to the idea that everything that stands will eventually lie down, just as everything that lies down will eventually stand up. The people who possess this knowledge have always lived in places where water abounds unceasingly: the Nile in Pharaonic Egypt. The Ampha and the Omega are just two expressions referring to a single thing, eternity, in the absence of any material element. The world of the immortals evoked in the countless stories of the Mvet (elevation of the spirit) is simply the very heart of typically spiritual life. The mutations that can be observed around the month of February, from 28 to 29 days, and the days of the year, from 365 to 366 days, express the idea of permanent resurrection. The triad of the sun in the sky effectively reflects this theological thinking. The pyramid is the symbol of eternal life.

It is important to note that the diagonal, triangular and fractal arrangement of the cordophones shows both an ascending and descending evolution, just like the quadrilaterals of the wave cycle. When the instrument is played by a master "mbom mvet" storyteller, he touches certain cordophones before others, which relates to the annual cycle of the months of the year. In other words, the universe, or at least the elements that make it up, live to the rhythm of the manipulation of a Divine being, who is its author. The dance is no more and no less than the expression of a vibrant homage paid to this Supreme Being. Isn't it said that it was Eyo'o who transmitted the Mvet to the warrior Ekang Oyono Ada Ngoine?

6-) The wave cycle and the Jewish candlesticks

The Jewish candlestick made up of 7 branches is called Menorah, whose etymological meaning refers to its origin inside a flame. In reality, it refers to the idea of luminosity, brilliance, resplendence, etc., in short to something that arouses administration. There is also the 9-branched candlestick called Hanukkiah, which also carries a flame. These two numbers are not at all foreign to us, as their respective numerical triplets are: 3+1+3=7 and 3+4+3=10, and 4+1+4=9 and

4+5+4=13. The main branch of the first is the 4th, while that of the second is the 5th. The luminous array of the Menorah stagnates at 3x5=15 and its numerical condensation is 27, while the Hanukkiah stagnates at 4x7=28 and knows the number 52 as the sum of the arithmetical condensation.

Naturally, we need to take them out of the calendar via the wave cycle, but to do that we first need to convert the latter into a purely geometric vision. Let's look at the first wave cycle below:

The Wave Cycle

2.		4		6				9		11
	*3		*5		*7	*8			*10	*12
1.	**2**	**3.**	**4.**	**5.**		**6.**		**7.**		
	*3		*5		*7	*8			*10	*12
2.		4		6				9		11

Source: NNANG EBANE Sosthene Tresor

We should always bear in mind that there is a period of stagnation between the months of July and August, which corresponds caricaturally to a horizontal line segment. According to a Maimonides author, the letter 'Y' is inscribed in the heart of the Menorah or seven-branched candlestick, but this needs to be demonstrated in a practical way. To achieve this, a geometric image is proposed:

The Wave Cycle

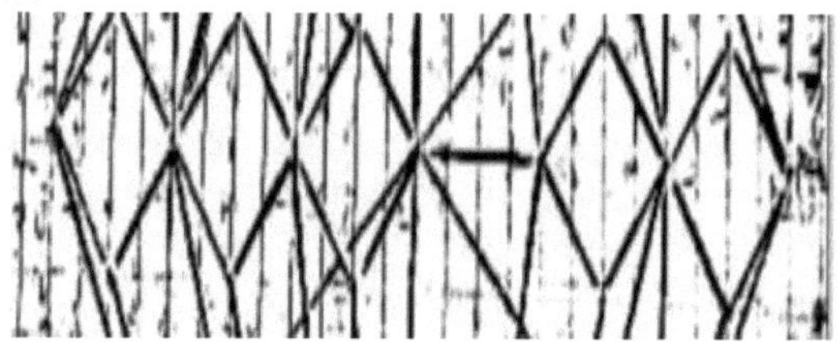

Source: NNANG EBANE Sosthene Tresor

If we look very carefully at the wave cycle, we can see that the vertical squares occupy the same position as in the luminous paintings. Also, that the period of stagnation unifies three (3) squares on the left-hand side against two (2) on the right-hand side. If we consider the horizontal segment of the right and divide the square into two equal parts, we will obtain the shape of the letter Y. This refers to the name Yawhe, meaning God. From the above, we understand that God is the heart of the universe.

To work out the number of candlesticks on this axis, we need to count the number of sides each square comprises, on either side of the stagnation period, which is in fact the main branch. On the left-hand side, there are 3 squares made up of 12 sides, i.e. 3x4=12. If we add the main branch, we get: (3x4)+1=12+1=13. On the other side, there are 2 squares made up of 8 sides, i.e. 2x4=8. If we add the main

160

branch, we get (2x4)+1=8+1=9. Against all expectations, we discover that the undulatory cycle is made up of two main candlesticks: the 13-arm candlestick and the 9-arm candlestick. The first runs from January to July, while the second runs from August to December.

In reality, the Menorah or 7-branch candlestick is an integral part of the 13-branch candlestick, as it constitutes its main branch. It must be said that six (6) branches are therefore extracted from it, hence the 7cm square reveals the number 6 as a numerical value within it. The sum of the candlesticks on the undulatory axis, 13 branches and 9 branches, adds up to 22, the number of letters in the Hebrew alphabet. It should be pointed out in passing that the number 27, resulting from the numerical condensation of the number 7, offers an extended view of this alphabet by adding the 5 sophists. In the light of the foregoing, we can see that the second wave cycle unifies each number of branches with the main ones, numerically speaking:

The Wave Cycle

2	4	6	8	10	12	
*3	*5	*7	*9	*11	*13	
1.	2	3.	4.	5.	6.	7.
*3	*5	*7	*9	*11	*13	
2	4	6	8	10	12	

Source: NNANG EBANE Sosthene Tresor

The introduction of candlesticks provides a perfect understanding of the notion of the numerical triplet, in terms of the number of branches they have. Let's consider only the numbers in black and those in red. If a candlestick has 3 branches, then the 2nd is its main branch. In short, the figures and numbers in black symbolise the main branches, while those in red designate those that play a balancing role.

The number 12, blue at the top and bottom, is an essential indicator for determining the complete number of branches in the new candlestick obtained. In fact, we can see that the number 13 is located in the centre of the branches, and we can see that T (n)=n+1+n equals T (12)=12+1+12=25 and T(n)=n+(n+1)+n equals T (12)=12+(12+1)+12=12+13+12=37. So it now has 25 branches, and the number 13 symbolises its main branch.

The geometrical form corresponding to the wave values is as follows:

The Wave Cycle

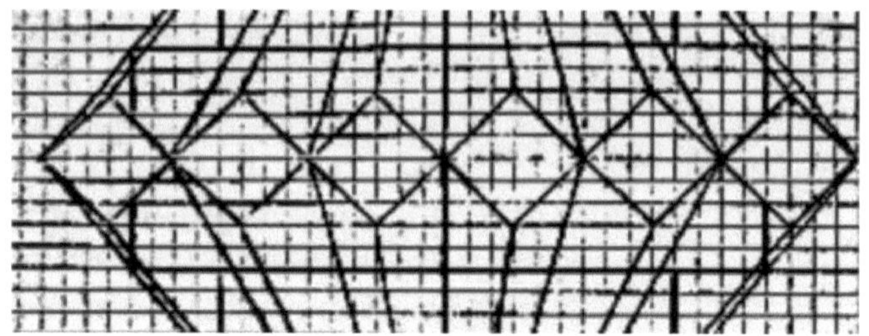

Excluding the stagnation seen between July and August, we get 6 vertical edges, by way of an undulatory cycle. The first cycle offers 5 edges, while the second is found at number 6, whose apex is 11. The numbers 5 and 6 represent the undulatory values of the 7 cm square. In view of the above, the calendar is a symbolic take on the number 7. Each square is made up of 4 sides, and there are 6 of them, so 4x6=24.

7-) The geometric perception of the calendar

Having presented the 7 axis of the calendar and its undulatory values, it is now a question of extracting its integral image. It was shown above that the number 7 has the number 4 as its centre of arithmetical symmetry, which coincides with the month of July (7) through these same numbers as the centre of the calendar from an undulatory perception. The height starting from this centre will be sectioned into four parts relating to the numbers 28, 29, 30 and 31, referring to the different months of the year. The number 28 will then form the vertex of this geometric figure, while at the same time forming its centre at the base. The following is a good illustration of what has just been said:

The time pyramid

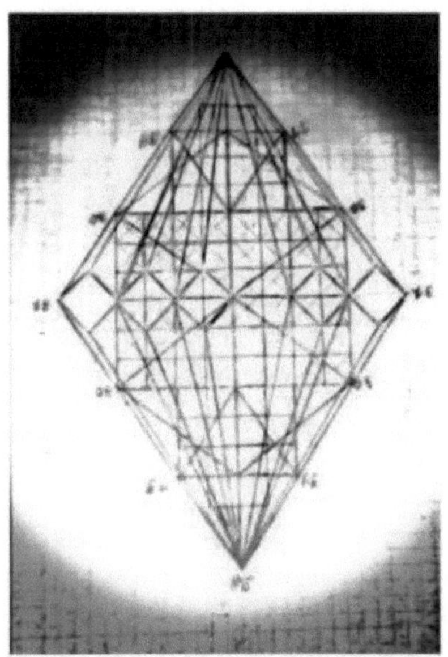

The resulting geometric shape is quite simply magnificent, the undulating centre seeming more imposing than the triangles extending from the centre towards both the heavens and the earth. What's more, another tripartite geometric shape is revealed within it. The quadrilateral in the centre is larger than the other two, which are located on the upper and lower sides of the same size. If we recall the configuration of the calabashes in the Mvet Ekang instrument, we can see that the largest is sometimes situated between two others of the same proportion. Similarly, the Jewish candlesticks, the Menorah and the Hanukkiah, sometimes have the main branch higher than the surrounding branches. The Sacred Harp fits in with this logic. The presence of the number 9 along the length of this rectangle in a square socket reveals the 8th diagonal as its centre of symmetry, giving it the base of a diagonometrical and fractal triangle. The calendar is therefore a pyramid bringing together all these religious instruments in a single socket. So the numerical triplet is said to be the expression of multiplicity within unity.

8-) The luminous panel

Very often, the Mvet Ekang instrument appears to have just one half of a central calabash below the cordophones, so we're going to look at the quadrilateral in the centre. This luminous rectangle has a very special feature, in that it combines the 7-branch candlestick and the 9-branch candlestick. The quadrilateral below will therefore be analysed in detail:

Figure: luminous panel9/4 and 7/3

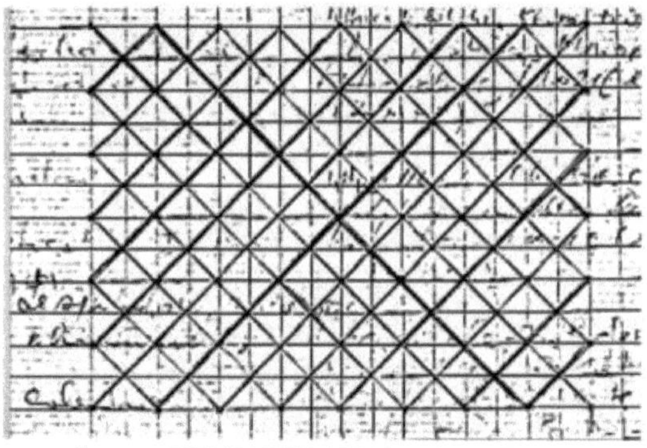

Source:NNANGEBANESostheneTresor

Horizontal reading: (7x6)+(8x5)=42+40=82

Vertical reading: (6x7)+(5x8)=42+40=82

Diagonal reading: 2+4+6+8+10+11+11+10+8+6+4+2=82

The triplet of the triangular reading is 30-22-30. It should be noted that the number 22 indicates the stagnant value of the triangle contained within this painting, yet the same number refers to the upper and lower readings of the diagonals contained within a square whose sides measure 7 cm in length. The number 7 is once again revealed as the centre of the universe.

The seven branches of the candlestick, according to a more naturalistic understanding, define the growth of a given tree. However, it is recognised that some trees grow according to seasons, which is where the notion of time comes in. So the geometrical and arithmetical configuration of the calendar is not at all surprising. The same applies to the Mvet Ekang, whose half calabashes are the fruit of the calabash tree. The growth of natural elements is made possible by the geometric view through the diagonals, which are capable of deforming the structure of a quadrilateral. When one of the diagonals is longer than the other, the quadrilateral automatically gains in volume, compared to when they are the same length. Smallness and size are two characteristics of a single element, which is why the cordophones of Ngombi and Mvet Oyeng are both ascending and descending. Similarly, the calendar counts months of 28, 30 and 31 days in a single year. Mathematics is not just about the simple manipulation of numbers and shapes, but rather about our inner unity with nature. It is an extreme knowledge of being under various natures, thus grouping all the sciences into a single nomenclature.

9-) Triplets 365 and 366

a-) paralleling triplets

Although the numerical triplet refers mainly to the triangle as a geometric figure built around the alignment of three (3) points, it also defines a certain evolution. Triplet 365 evokes an ascent from 3 to 6, then a regression from 6 to 5. On the other hand, triplet 366 refers to growth from 3 to 6, then stagnation from 6 to 6, demonstrating that there can be no evolution without a period of rest. Both growth

164

and decay experience a resting period, which allows them to either resume growth or decay. Here, rest seems to refer to a certain focus of regeneration, reconstruction and even reconstitution of consciousness. In this way, the question of the eternal life of memory is transposed to the very heart of figures and numbers. There is always a living organism within another, which it uses just to discover itself, and which is thus dependent on it. The division of arithmetical triplets (03) into zones of two (02) limits finally reveals the unity of four (4) points, either (3;6) and (6;5) or (3;6) and (6;6). Hence, the triangle known as the life-giving, is always associated with the square known as the life-giving. We count 4 fingers on our hands, but there are 3 intervals between these fingers, which is why the number evokes the invisible, unlike the first, which is matter. The pyramid therefore has a symbolic value as a practical manifestation of life. As the triangle is both inside and outside the square, it expresses its original spiritual nature. The triangle thus possesses a life outside the square, which explains why the pyramid itself is the work of the spirit or the invisible.

b-) triples 365, 366 and diagonals

The parallel relationship of the triplets has been followed up to this point, because the first needs to be superimposed on the second in the same way as the division. The first will therefore be located at the top level, while the second will be at the bottom level, after which the different readings will be brought out diagonally. It is therefore possible to divide the triplets into two blocks: 36/36 and 65/66.

The diagonometric core (36X) of the axis

3   6

*   *

3   6

Source: NNANG EBANE Sosthene Tresor

The addition ratio reveals the number 9 as the result of the operations 3+6=9 and 6+3=9, then 3+6=9 and 6+3=9, hence the number 9999, or 4^9=36. The number 9999 can be split into two large groups, a triplet 999 and a quadruple 9999, hence the revelation of the number 7, taken as the construction axis of the Gregorian calendar. However, the triangular, parallel, fractal and diagonometric arrangement of this quadruple reveals the number 8, as the interval value of the number 9. So we have 8 upper perforations, 8 cordophones, lower perforations, hence 3x8=24 (Ngombi). The clean result (9) is not far from the number 24, symbolising the length of day and night. The square with a length of 9 perforations has wave values 7 and 8, so 7+8+9=24. The mathematical logic associated with the notion of an arithmetical triplet presents two possibilities: the first relates to the trilateral figure (triangle) through the sum of three aligned points. The second is built around the square and the diagonals as the regression of a given number n in two dimensions noted n-2 and n-1. Hence the number 9, where n decreases to 7 (n-2), then to 8 (n-1). T(n)=(n-2)+(n-1) equals T (9)=(9-2)+(9-1)=7+8. Then the heart changes digits:

The diagonometric core (36X) of the axis

9   9

*   *

9   9

Source: NNANG EBANE Sosthene Tresor

Quite logically, the 9-perforation square from which the Ngombi/Ngoma, or Sacred Harp, is drawn, appears in all its splendour. The number 36 would therefore designate the perimeter of the 4x9 square. The number 9 is thus associated with the revolution of time - don't we say that the duration of a pregnancy is 9 months? The number therefore evokes the idea of evacuation, vomiting, exteriorisation, the manifestation of life, taking place over time. The symbolic vision of the Sacred Harp therefore relates to the discovery of the inner self. Spirituality is thus linked to universal practical facts, experienced by all, and verifiable by all.

Let's now look at the second half of the numerical triplets we've superimposed:

The diagonometric core (36X) of the axis

6   5

*   *

6   6

Source: NNANG EBANE Sosthene Tresor

We can see that the design has 3 identical numbers, and another that is different. The number 666 can be seen in the triangular layout of the square, which is 7 perforations long. According to the Sacred Harp model, there are 6 upper perforations, 6 cordophones and 6 lower perforations, i.e. 3x6=18. The undulatory values of the number 7 are 5 and 6, so 5+6+7=18, hence the number 666 is only a homogeneous perception of the triplet 567. The number 18 is the sum of 9+9, which explains the palpable correlation between the number 7 and the number 9. Both are part of the numerical triplets: 3+1+3=7 and 4+1+4=9, a symbolic perception of the Menorah (7-branch candlestick) and the Hanukkiah (9-branch candlestick). Hence the Gregorian calendar is seen as a concrete reality of the Jewish candlesticks.

A diagonometric reading reveals the sums 11=5+6 and 6+5=11, and 12=6+6 and 12=6+6. Furthermore, 11+11=22 and 12+12=24. Remembering the first temporal axis, it has been shown that not only are the 12 months of the year built around the number 7, but also that the same axis is the sum of the unity of the 13-branch candlesticks, of which the number 7 symbolises the main one, and the 9-branch candlestick, so 9 branches plus 13 branches equals 22 branches. In view of the above, the candlesticks refer to the passing of time, which regenerates, while accompanying the work of procreation.

166

The diagonometric core (36X) of the axis

22   24

*   *

24   22

Source: NNANG EBANE Sosthene Tresor

The number 12 is a globalising value, while the number 11 is the numerical value of its interval, resulting in an increase in each of the values 11+11=22 and 12+12=24. However, we can see that the number 22 is not the real interval value of the number 24, so we add a unit to it, giving us 23 and 24. The body is therefore symbolised by the number 24, whereas the number 22 refers to the spirit. In the light of the above, the diagonals seem to show that we must always have a very particular eye on the content of the spirit and not on that of the pulpit. One of the diagonals, 24, is longer than the 22, indicating that there is always a balance of power between carnal and spiritual desires. The diagonals therefore symbolise man's heart, since they are inside the square (body). Life is therefore internal and not external to the human body, which is no more and no less than a simple mirror reflecting inner thoughts. The time thus evoked by calendars is not always relative to the seasons, but also to the perpetual struggle between matter and spirit. Hence, it would be a mistake to claim that the Ekang Mvet, the Punu Mask, the Songo, the Ngombi, and the Jewish Candlesticks are civilisational realities, completely alien to our times. We are talking here about the unconditional preservation of memory, which is in reality the very expression of time, and therefore of eternity. I have always been convinced that the Ancestors are open to all those who are interested in their history. It cannot sink into the shadows of oblivion, because the rites of land possession were celebrated in advance, to perpetuate this knowledge. Time doesn't shout from the rooftops to make itself understood, but it brings everyone into agreement when it comes. The memory of oblivion is a reactualised birth that merges with the present, with the aim of filling the void that rendered us so powerless in the astonished eyes of our detractors.

b-) triples 365, 366, diagonals and axes of symmetry

By superimposing the triplet 365 on the triplet 366, linking the upper number 3 to the lower number 6, and the lower number 3 to the upper number 5, we obtain the diagonal numbers 3+6=9 and 3+5=8. However, the central ratio is 12=6+6, 12 higher and 12 lower. The numbers (9,9) and (8,8) symbolise the diagonals, but the numbers (6,6) refer to the axis of symmetry, revealing the very nature of a quadrilateral. There are two (2) diagonals and one (1) axis of symmetry, each with two (2) extremities, so the numbers 3 and 6 are the numerical symbols of a quadrilateral. The number 9=3+6, summarises the unity of the diagonals with an axis of symmetry.

It has been shown above that the number 8 is an integral value of the number 9 and is also the triplet value of the triangulation of the 9-perforation square, the corresponding triplet of which is 888 or 3x8=24. In addition, the number 12 represents the degree of diagonometric stagnation of the 8-perforation square. The number is of a numerical value greater than the calendar axis of 7, 7+1=8, so the number symbolises renewal. It thus embodies the passage from December to January, symbolising the 13th month.

The number 9 is the expression of mobility, because it corresponds to the number 28 (length of days in February), through its luminous table whose stagnation is 7777 or 4x7=28. According to the luminous table of 9, the triplet 888 (i.e. 3x8=24) is located between the quadruple 7777 whose practical view is 7878787, so 7+8+9=24.

This arithmetical correspondence is cleverly inscribed in the Ngoma, Ngombi, or Sacred Harp of Gabon. Furthermore, the number 28 plus one (1) unit refers to the number 29. This number is equal to the sum of (3+6+5)+(3+6+6)=14+15=29. Similarly, 365+1=366 symbolises the numerical value of the interval (invisible) and the practical arithmetical value (physical) respectively.

February has fewer days than the other months, but it embodies mobility or life through the variation in the number of its days, while the other months remain immobile. In the same vein, the stringophones of stringed instruments represent the only flexible part. Life is organised around a column from which numerous ramifications emerge to infinity. So the calendar is not just about reading time as it passes, but also about reading time as it never changes. There is always an origin anterior to all existence. Understanding this existence certainly leads to this unknown origin. Genealogy is quite simply a manifest recognition that a people belongs to an invisible creator.

CONCLUSION

The appreciation of an object is most often conditioned by the framework in which it is traditionally presented, because it carries in the forefront a major idea that is intended to be generalized. The sharing of common knowledge and its preservation by a group of individuals is nothing to complain about, but what is to complain about is the formal prohibition of its study outside the initial field of interaction. It's like a strange phenomenon that sometimes appears out of nowhere and touches the observer in a more intimate way, pushing him to pursue his understanding of the object beyond the old frontier limits. This departure from the story is unfortunately seen as a high betrayal, on the part of the elders who want to preserve a common heritage, but according to the original model. This approach certainly seems more logical in a society where individuals no longer know where to turn in order to reappropriate ancestral knowledge that has long been labelled diabolical. It is perfectly natural to start from this knowledge to build a society in the image of a strain of its own. Shouldn't the spiritual elevation to which the term 'Mvet' refers find meaning within this new vision? The quest for truth is a path that requires a truly conscientious approach aimed at satisfying a personal thirst, which we have no wish to allow to be obscured by an external will. Sacred instruments in themselves are discourses to be listened to, just as they are manipulated by men. Just as oppressed peoples must define themselves on their own terms. Spirituality also requires a certain ability to personally merge with the object in order to understand it. Unity with the invisible is supposed to be a personal preoccupation, which may later be accompanied by singular or collective external help. The expected result tends to unify the traditional arts, even though they do not belong to the same ethnic group, which raises questions about the dissemination of a common knowledge whose origins are not unanimously agreed? The Fangs, the Punu and the Hebrews occupy different geographical areas, but their arts are linked to Pharaonic Egypt through the pyramid? This region of the world is the epicentre of spiritual and scientific knowledge. The study of the sacred arts tends to define them as brother peoples, because they share a common ancestral heritage, beyond the oral discourse that places them on opposite sides of the world. This new appreciation of the sacred arts is in line with the thoughts of the illustrious Gabonese writer Fidele OKUE NgOU, who said: *"I like to think that many young people from your ðënëzauon will have the vocation to seek out the treasures of our culture at the end of their university studies. They are still buried in the humus of our villages"* The science of mathematics long presented as a school discipline and as a reality external to the African peoples, reveals itself against all odds within their traditional spiritual arts. However, the spiritual aspect is more prominent than the mathematical one, which is perhaps kept silent by the hateful action of colonisation, by simple ignorance, or by the fact that the mind is an unconditional dominant of matter? That is to say, to the essential, which is the invisible. The cordophones are the only flexible part of the Mvet Ekang and Ngombi, located at their centre, just as the human soul is inside the body. The Egyptian context adds a particular flavour, through the sound that symbolises the waters of the Nile, i.e. the vivifier giving life to the vivifier (pyramid). Time is not just about the hours that pass like the waters of the river, but also about the manifest expression of life, through movement within an immobile structure. The month of February is rightly characterised by mobility through the variation in the number of days that make up the month - 28 or 29 - in contrast to the stagnation of other months. The sacred African arts reveal life as being contained within a structure. The structure is not life, but it contains it. Is this how life was originally formed? The number 1, which refers to the evolution of figures and numbers at intervals, indicates that the number 28 is not only earlier, but also superior to the number 29, which is its mutation. So the latter cannot be superior to the former according to the other of creation, because it is based on the former. The child cannot be defined as dominant over its parents? Respect for the invisible is no longer at the height of its glory in this modern age; the pulpit is taking on enormous power over the mind, hence the perdition of the human race. Modernity is built on approximations that do not guarantee the wellbeing of the human soul; on the contrary, it is underpinned by other immoral values such as parental abandonment, divorce, denial of identity, and so on. Genealogy or respect for elders remains the backbone that links people to the invisible, of which the palm branch (a time indicator) remains a very poignant symbol. The twin possibilities of the ascension and regression of cordophones in a single take perfectly illustrate the unity of generations. The Alpha and Omega, which are religious adjectives attributed to God, are obviously to be found here. Is this not reason enough to demand the return of our civilisational memories held captive in the poorly varnished coffins of our museums?

SOURCES AND BIBLIOGRAPHY

1-) Websites

a-) culture

https://www.amazon.fr>calendar-multicomore-43c...

http s://www j epense. org>le-triangle-et-son- sypbolisme

https://www.fr.m.wikipedia.org>wiki>menorah/2020(8 december)

https://www.fr.m.wikipedia.org>wiki>menorah-symbol-of-judaism/2021 (May 27)

https://www.laportedubonheur.com>history-and-significance...

http s: //www.fr.m. wikip edia. org>wiki>mvett

https://www.cursus.edu>understanding-the-transmission-of-oral-knowledge-through-the-mvet/(2023)

https://www.wikiwand.com>the-people-fang

https://www.art-masque-africain.com>the-punu-mask

https://www.corep.fr>definition-schedule

https://www.choisir.ch/symboliques-des-lettres-hebraiques/2018 (10 January)

https://www.numeration-hebraique

http s: //www.anthio cus.over-blog.com>Tav

https://www.laportedubonheur.com>history-and-significance-of-the-david-star

https://www.mummies2pyramid.com>image

https://www.etre-nature.fr>alphabet-hebrew

https://www.acoursdhebreu.com>alphabet-hebrew

https ://www.clubawale.com>songo

https://www.bantoozone.org>ngoma-harpe-fang-music-gabon

https://www.mon-gabon.com>province-of-woleu-ntem

https://www.culturebene.com>the-origins-of-songo-this-mythical-game-of-the-ekang-people

https://www.lepratiquedugabon.com>the-masks-of-gabon

https://www.holyart.fr>blog>religious-article/the-menorah-a-sept-branches-(2020)

b-) mathematics

https://www.fr.m.wikipedia.org>wiki>the-pascal-triangle

https://www.paramaths.fr>pascal-triangle/nicolas-masset/3(june 2022)

https://www.mathcoaching.com>card>recognising-and-describing-a-triangle-rectangle

https://www.fr.m.wikipedia.org>wiki>geometrie-sacree

https://www.fr.m.wikipedia.org>wiki>fractals/2021(July 8)

https://www.larousse.fr>thematics

https://www.educastream.com>the-rectangle

https://www.clg-monnet-bruiis-versailles.fr>the-quadrilateres-convex-crossings-concave

https://www.centre-scoences.org>the-calendar

https://www.nationalgeographic.fr>the-phases-of-the-moon

2-) Bibliography

a-) culture

Fidele OKUE NgOU, Elements typiques de la culture fang, Mitzic, 2003 p.91

b-) mathematics

Collection fractales mathematiques tles A1et B, Bordas, Paris, 1992 p.2

Collection inter africaine de mathematiques, resoudre les problemes geometriques, EDICEF 1995 p.7

Information about l'auteur

nnangebanes@gmail.com

Born on 30/09/1989 in Okala, a small village in the province of Woleu-Ntem in Gabon. Studied at the Omar Bongo University in Libreville in the Department of Anglephone Studies, where he obtained a degree. He currently works as a security agent for Vigile Services Protection (VSP), a company formerly known as Gabon Poste. A lover of ancestral life, he is trying to retrace its steps through this work, which is the fruit of a strong request from friends, who found him quite well established on the subject. The work is based exclusively on his observations of the sacred arts: the Menorah, the Mvet Ekang, the Hanukkia, the Punu mask, the Sacred Harp (Ngombi or Ngoma), and the Songo, for which the calendar serves as a point of reference. He makes it clear, however, that he is not a mathematician by training, but rather that he lets himself be carried away by the intoxicating scientific universe of civilisational memories.

Buy your books fast and straightforward online - at one of world's fastest growing online book stores! Environmentally sound due to Print-on-Demand technologies.

Buy your books online at
www.morebooks.shop

Kaufen Sie Ihre Bücher schnell und unkompliziert online – auf einer der am schnellsten wachsenden Buchhandelsplattformen weltweit! Dank Print-On-Demand umwelt- und ressourcenschonend produziert.

Bücher schneller online kaufen
www.morebooks.shop